Ernest Spon

Water Supply

The Present Practice of Sinking and Boring Wells

Ernest Spon

Water Supply
The Present Practice of Sinking and Boring Wells

ISBN/EAN: 9783743401549

Manufactured in Europe, USA, Canada, Australia, Japa

Cover: Foto ©berggeist007 / pixelio.de

Manufactured and distributed by brebook publishing software (www.brebook.com)

Ernest Spon

Water Supply

WATER SUPPLY.

THE PRESENT PRACTICE

OF

SINKING AND BORING WELLS.

THE PRESENT PRACTICE

OF

SINKING AND BORING WELLS.

WITH

GEOLOGICAL CONSIDERATIONS AND EXAMPLES OF
WELLS EXECUTED.

BY

ERNEST SPON,

ASSOC. M. INST. C.E.; MEMBER OF THE SOCIETY OF ENGINEERS; OF THE FRANKLIN
INSTITUTE; AND OF THE GEOLOGISTS' ASSOCIATION.

SECOND EDITION.

E. & F. N. SPON, 125, STRAND, LONDON.

NEW YORK; 35, MURRAY STREET.

1885.

PREFACE.

A LONG preface is not necessary to explain the purport of this work. Suffice it to say that it is meant to afford a short but comprehensive account of modern practice in obtaining water by means of wells, derived from personal experience as well as from that of the highest authorities.

The importance of this means of water supply, instead of diminishing, becomes enhanced with every fresh development of municipal or industrial life, and commands increased confidence with the growth of the science upon which it is based.

In the first edition, attention was drawn to the fact that a great deal of the irregularity in the action of wells, and the consequent distrust with which they are regarded by many, is attributable either to improper situation, or to the hap-hazard manner in which the search for underground water is too frequently conducted. As regards the first cause, extreme caution is necessary in the choice of situations, for wells, and a sound geological knowledge of the country in which the attempt is to be made should precede sinking or boring for this purpose, otherwise much useless expense may be incurred without success. Indeed the power of indicating those places where wells may in all probability be successfully established, is one of the chief practical applications of geology to the useful purposes of life.

The subject-matter of the following pages is divided into

chapters which treat of geological considerations, the Jurassic strata, the new red sandstone, well sinking, well boring, tube wells, well boring at great depths, and examples of wells executed, and of localities supplied respectively, with tables and miscellaneous information.

As heretofore, I have to acknowledge my indebtedness to Professor Prestwich, Messrs. S. Baker and Son, and T. Docwra and Son, and to express my acknowledgments to A. Harston, Esq., of London, M. Leon Dru, Paris, with the many other correspondents, from all parts of the globe, who have sent me local information and sections. For such communications I have always a warm welcome, and I hope that the issue of the present edition, so long delayed by pressure of active professional work, will cause their number to increase and multiply.

ERNEST SPON.

7, IDOL LANE, LONDON.
  December, 1884.

CONTENTS.

CHAPTER I.

CHAPTER VII.

CHAPTER VIII.

CHAPTER IX.

SINKING AND BORING WELLS.

CHAPTER I.

GEOLOGICAL CONSIDERATIONS.

NEARLY every civil engineer is familiar with the fact that certain porous soils, such as sand or gravel, absorb water with rapidity, and that the ground composed of them soon dries up after showers. If a well be sunk in such soils, we often penetrate to considerable depths before we meet with water; but this is usually found on our approaching some lower part of the porous formation where it rests on an impervious bed; for here the water, unable to make its way downwards in a direct line, accumulates as in a reservoir, and is ready to ooze out into any opening which may be made, in the same manner as we see the salt water filtrate into and fill any hollow which we dig in the sands of the shore at low tide. A spring, then, is the lowest point or lip of an underground reservoir of water in the stratification. A well, therefore, sunk in such strata will most probably furnish, besides the volume of the spring, an additional supply of water, inasmuch as it may give access to the main body of the reservoir.

The transmission of water through a porous medium being so rapid, we may easily understand why springs are thrown out on the side of a hill, where the upper set of strata consist of chalk, sand, and other permeable substances, whilst those lying beneath are composed of clay or other retentive soils. The only difficulty, indeed, is to explain why the water does not ooze out everywhere along the line of junction of the two formations, so as to form one continuous land-soak, instead of

B

a few springs only, and these oftentimes far distant from each other. The principal cause of such a concentration of the waters at a few points is, first, the existence of inequalities in the upper surface of the impermeable stratum, which lead the water, as valleys do on the external surface of a country, into certain low levels and channels; and secondly, the frequency of rents and fissures, which act as natural drains. That the generality of springs owe their supply to the atmosphere is evident from this, that they vary in the different seasons of the year, becoming languid or entirely ceasing to flow after long droughts, and being again replenished after a continuance of rain. Many of them are probably indebted for the constancy and uniformity of their volume to the great extent of the subterranean reservoirs with which they communicate, and the time required for these to empty themselves by percolation. Such a gradual and regulated discharge is exhibited, though in a less perfect degree, in all great lakes, for these are not sensibly affected in their levels by a sudden shower, but are only slightly raised, and their channels of efflux, instead of being swollen suddenly like the bed of a torrent, carry off the surplus water gradually.

An Artesian well, so called from the province of Artois, in France, is a shaft sunk or bored though impermeable strata, until a water-bearing stratum is tapped, when the water is forced upwards by the hydrostatic pressure due to the superior level at which the rain-water was received. The term Artesian was originally only applied to wells which overflowed, but nearly all deep wells are now so called, without reference to their water-level, if they have bore-holes.

Among the causes of the failure of Artesian wells, we may mention those numerous rents and faults which abound in some rocks, and the deep ravines and valleys by which many countries are traversed; for when these natural lines of drainage exist, there remains only a small quantity of water to escape by artificial issues. We are also liable to be baffled by the great thickness either of porous or impervious strata, or by the dip of the beds, which may carry off the waters from

adjoining high lands to some trough in an opposite direction—
as when the borings are made at the foot of an escarpment
where the strata incline inwards, or in a direction opposite to
the face of the cliffs.

As instances of the way in which the character of the strata
may influence the water-bearing capacity of any given locality,
we give the following examples, taken from Latham. Fig. 1

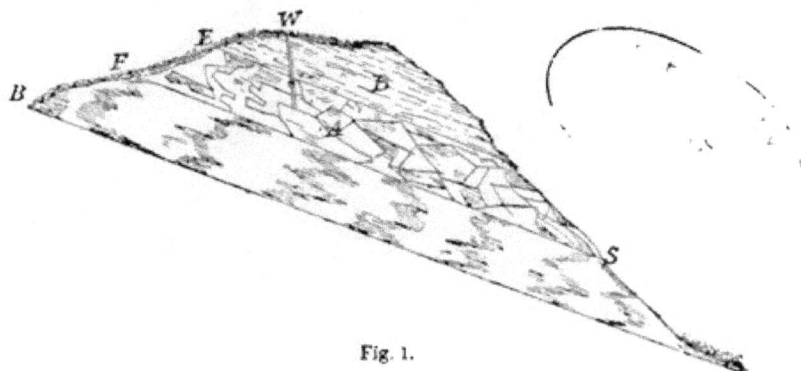

Fig. 1.

illustrates the causes which sometimes conduce to a limited
supply of water in Artesian wells. Rain descending on the
outcrop E F of the porous stratum A, which lies between the
impervious strata B B, will make its appearance in the form of
a spring at S; but such spring will not yield any great

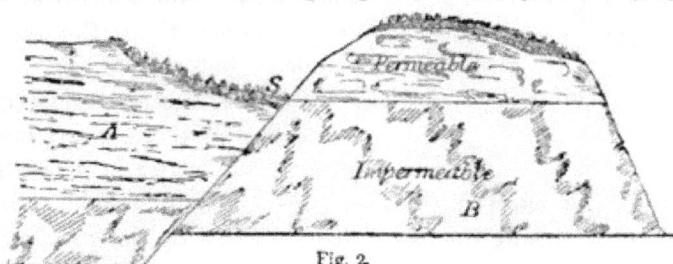

Fig. 2.

quantity of water, as the area E F, which receives the rainfall,
is limited in its extent. A well sunk at W, in a stratum of the
above description, would not be likely to furnish a large supply
of water, if any. The effect of a fault is shown in Fig. 2. A
spring will in all probability make its appearance at the point

S, and give large quantities of water, as the whole body of water flowing through the porous strata A is intercepted by being thrown against the impermeable stratum B. Permeable rock intersected by a dyke and overlying an impermeable stratum is seen in Fig. 3. The water flowing through A, if

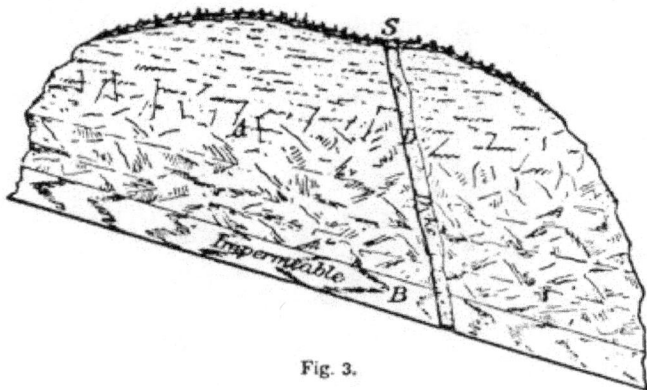

Fig. 3.

intersected by a dyke D, will appear at S in the form of a spring, and if the area of A is of large extent, then the spring S will be very copious. As to the depth necessary to bore certain wells in a case similar to Fig. 4, owing to the fault, a well sunk at A would require to be sunk deeper than the well B, although both wells derive their supply from the same description of strata. If there is any inclination in the water-bearing strata,

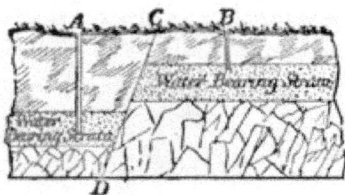

Fig. 4.

or if there is a current of water only in one direction, then one of the wells would prove a failure owing to the proximity of the fault, while the other would furnish an abundant supply of water.

It should be borne in mind that there are two primary geological conditions upon which the quantity of water that may be supplied to the water-bearing strata depends; they are, the extent of superficial area they present, by which the quantity of rain-water received on their surface in any given time is determined; and the character and thickness of the strata, as by this the proportion of water that can be absorbed, and the quantity which the whole volume of the permeable strata can transmit, is regulated. The operation of these general principles will constantly vary in accordance with local phenomena, all of which must, in each separate case, be taken into consideration.

The mere distance of hills or mountains need not discourage us from making trials; for the waters which fall on these higher lands readily penetrate to great depths through highly-inclined or vertical strata, or through the fissures of shattered rocks; and after flowing for a great distance, must often reascend and be brought up again by other fissures, so as to approach the surface in the lower country. Here they may be concealed beneath a covering of undisturbed horizontal beds, which it may be necessary to pierce in order to reach them. The course of water flowing underground is not strictly analogous to that of rivers on the surface, there being, in the one case, a constant descent from a higher to a lower level from the source of the stream to the sea; whereas, in the other, the water may at one time sink far below the level of the ocean, and afterwards rise again high above it.

✗ For the purposes under consideration, we may range the various strata of which the outer crust of the earth is composed under four heads, namely: 1, drift; 2, alluvion; 3, the tertiary and secondary beds, composed of loose, arenaceous and permeable strata, impervious, argillaceous and marly strata, and thick strata of compact rock, more or less broken up by fissures, as the Norwich red and coralline crag, the Molasse sandstones, the Bagshot sands, the London clay, and the Woolwich beds, in the tertiary division; and the chalk, chalk marl, gault, the greensands, the Wealden clay, and the Hastings sand, the oolites, the lias, the Rhætic beds, and Keuper, and the new red sandstone, in the secondary division; and 4, the primary beds, as the magnesian limestone, the lower red sand, and the coal measures, which consist mainly of alternating beds of sandstones and shales with coal.

The first of these divisions, the drift, consisting mainly of sand and gravel, having been formed by the action of flowing water, is very irregular in thickness, and exists frequently in detached masses. This irregularity is due to inequalities of the surface at the period when the drift was brought down. Hollows then existing would often be filled up, while either none was deposited on level surfaces, or, if deposited, was subsequently removed by denudation. Hence we cannot infer when boring through deposits of this character that the same, or nearly the same, thickness will be found at even a few yards' distance. In valleys this deposit may exist to a great depth, the slopes of hills are frequently covered with drift, which has either been arrested by the elevated surface or brought down from the upper portions of that surface by the action of rain. In the former case the deposits will probably consist of gravel, and in the latter of the same elements as the hill itself.

The permeability of such beds will, of course, depend wholly upon the nature of the deposit. Some rocks produce deposits through which water percolates readily, while others allow a passage only through such fissures as may exist. Sand and gravel constitute an extremely absorbent medium, while an argillaceous deposit may be wholly impervious. In mountainous districts, springs may often be found in the drift; their existence in such formations will, however, depend upon the position and character of the rock strata; thus, if the drift cover an elevated and extensive slope of a nature similar to that of the rocks by which it is formed, springs due to infiltration through this covering will certainly exist near the foot of the slope. Upon the opposite slope, the small spaces which exist between the different beds of rock receive these infiltrations directly, and serve to completely drain the deposit which, in the former case, is, on the contrary, saturated with water. If, however, the foliations or the joints of the rocks afford no issue to the water, whether such a circumstance be due to the character of their formation, or to the stopping up of the issues by the drift itself, these results will not be produced.

It will be obvious how, in this way, by passing under a mass of drift, the water descending from the top of hill slopes re-appears at their foot in the form of springs. If now we suppose these issues stopped, or covered by an impervious stratum of great thickness, and this stratum pierced by a boring, the water will ascend through this new outlet to a level above that of its original issue, in virtue of the head of water measured from the points at which the infiltration takes place to the point in which it is struck by the boring.

Alluvion, like drift, consists of fragments of various strata carried away and deposited by flowing water; it differs from the latter only in being more extensive and regular, and, generally, in being composed of elements brought from a great distance and having no analogy with the strata with which it is in contact. Usually it consists of sand, gravel, rolled pebbles, marls or clays. The older deposits often occupy very elevated districts, which they overlie throughout a large extent of surface. At the period when the large rivers were formed, the valleys were filled up with alluvial deposits, which at the present day are covered by vegetable soil, and a rich growth of plants, through which the water percolates more slowly than formerly. The permeability of these deposits allows the water to flow away subterraneously to a great distance from the points at which it enters. Springs are common in the alluvion, and more frequently than in the case of drift, they can be found by boring. As the surface, which is covered by the deposit, is extensive, the water circulates from a distance through permeable strata often overlaid by others that are impervious. If at a considerable distance from the points of infiltration, and at a lower level, a boring be put down, the water will ascend in the bore-hole in virtue of its tendency to place itself in equilibrium. Where the country is open and sparsely inhabited, the water from shallow wells sunk in alluvion is generally found to be good enough, and in sufficient quantity for domestic purposes.

The strata of the tertiary and secondary beds, especially the latter, are far more extensive than the preceding, and yield much larger quantities of water. The chalk is the great water-bearing stratum for the larger portion of the south of England. The water in it can be obtained either by means of ordinary shafts, or by Artesian wells bored sometimes to great depths, from which the water will frequently rise to the surface. It should be observed that water does not circulate through the chalk by general permeation of the mass, but through fissures. A rule given by some for the level at which water may be found in this stratum is, " Take the level of the highest source of supply, and that of the lowest to be found. The mean level will be the depth at which water will be found at any intermediate point, after allowing an inclination of at least 10 feet a mile." This rule will also apply to the greensand. This formation contains large quantities of water, which is more evenly distributed than in the chalk. The gault clay is interposed between the upper and the lower greensand, the latter of which also furnishes good supplies. In boring into the upper greensand, caution should be observed so as not to pierce the gault clay, because water which permeates through that system becomes either ferruginous, or contaminated by salts and other impurities.

The next strata in which water is found are the upper and inferior oolites, between which are the Kimmeridge and Oxford clays, which are separated by the coral rag. There are instances in which the Oxford clay is met with immediately below the Kimmeridge, rendering any attempt at boring useless, because the water in the Oxford clay is generally so impure as to be unfit for use. And with regard to finding water in the oolitic lime-stone, it is impossible to determine with any amount of precision the depth at which it may be reached, owing to the numerous faults which occur in the formation. It will therefore be neces-sary to employ the greatest care before proceeding with any borings. Lower down in the order are the upper lias, the marl-stone, the lower lias, and the new red sandstone. In the marl-stone, between the upper and lower beds of the lias, there may be found a large supply of water, but the level of this is as a rule too low to rise to the surface through a boring. It will be necessary to sink shafts in the ordinary way to reach it. In the new red sandstone, also, to find the water, borings must be made to a considerable depth, but when this formation exists a copious supply can be confidently anticipated, and when found the water is of excellent quality.

Every permeable stratum may yield water, and its ability to do this, and the quantity it can yield, depend upon its position and extent. When underlaid by an impervious stratum, it constitutes a reservoir of water from which a supply may be drawn by means of a sinking or a bore-hole. If the permeable stratum be also overlaid by an impervious stratum, the water will be under pressure and will ascend the bore-hole to a height that will depend on the height of the points of infiltration above the bottom of the bore-hole. The quantity to be obtained in such a case as we have already pointed out, will depend upon the extent of surface possessed by the outcrop of the permeable stratum. In searching for water under such conditions a careful examination of the geological features of the district must be made. Frequently an extended view of the surface of the district, such as may be obtained from an eminence, and a consideration of the particular configuration of that surface, will be sufficient to enable the practical eye to discover the various routes which are followed by the subterranean water, and to predicate with some degree of certainty that at a given point water will be found in abundance, or that no water at all exists at that point. To do this, it is sufficient to note the dip and the surfaces of the strata which are exposed to the rains. When these strata are nearly horizontal, water can penetrate them only through their fissures or pores; when, on the contrary, they lie at right angles, they absorb the larger portion of the water that falls upon their outcrop. When such strata are intercepted by valleys, numerous springs will exist. But if, instead of being intercepted, the strata rise around a common point, they form a kind of irregular basin, in the centre of which the water will accumulate. In this case the surface springs will be less numerous than when the strata are broken. But it is possible to obtain water under pressure in the lower portions of the basin, if the point at which the trial is made is situate below the outcrop.

The primary rocks afford generally but little water. Having

been subjected to violent convulsions, they are thrown into
every possible position and broken by numerous fissures; and
as no permeable stratum is interposed, as in the more recent
formations, no reservoir of water exists. In the unstratified
rocks, the water circulates in all directions through the fissures
that traverse them, and thus occupies no fixed level. It is also
impossible to discover by a surface examination where the
fissures may be struck by a boring. For purposes of water
supply, therefore, these rocks are of little importance. It must
be remarked here, however, that large quantities of water are
frequently met with in the magnesian limestone and the lower
red sand, which form the upper portion of the primary series.

Joseph Prestwich, jun., in his 'Geological Inquiry respecting
the Water-bearing Strata round London,' gives the following
valuable epitome of the geological conditions affecting the value
of water-bearing deposits; and although the illustrations are
confined to the Tertiary deposits, the same mode of inquiry
will apply with but little modification to any other formation.

The main points are—

The extent of the superficial area occupied by the water-
bearing deposit.

The lithological character and thickness of the water-bearing
deposit, and the extent of its underground range.

The position of the outcrop of the deposit, whether in valleys
or hills, and whether its outcrop is denuded, or covered with
any description of drift.

The general elevation of the country occupied by this outcrop
above the levels of the district in which it is proposed to sink
wells.

The quantity of rain which falls in the district under consideration, and whether, in addition, it receives any portion of
the drainage from adjoining tracts, when the strata are impermeable.

The disturbances which may affect the water-bearing strata,
and break their continuous character, as by this the subterranean flow of water would be impeded or prevented.

Extent of Superficial Area.

To proceed to the application of the questions in the particular instance of the lower tertiary strata. With regard to the first question, it is evident that a series of permeable strata, encased between two impermeable formations, can receive a supply of water at those points only, where they crop out and are exposed on the surface of the land. The primary conditions affecting the result depend upon the fall of rain in the district where the outcrop takes place; the quantity of rain-water which any permeable strata can gather being in the same ratio as their respective areas. If the mean annual fall in any district amounts to 24 inches, then each square mile will receive a daily average of 950,947 gallons of rain-water. It is therefore a matter of essential importance to ascertain, with as much accuracy as possible, the extent of exposed surface of any water-bearing deposit, so as to determine the maximum quantity of rain-water it is capable of receiving.

The surface formed by the outcropping of any deposit in a country of hill and valley is necessarily extremely limited, and it would be difficult to measure in the ordinary way. Prestwich therefore used another method, which seems to give results sufficiently accurate for the purpose. It is a plan borrowed from geographers, that of cutting out from a map, on paper of uniform thickness, and on a large scale, say one inch to the mile, and weighing the superficial area of each deposit. Knowing the weight of a square of 100 miles cut out of the same paper, it is easy to estimate roughly the area in square miles of any other surface, whatever may be its figure.

Mineral Character of the Formation.

The second question relates to the mineral character of the formation, and the effect it will have upon the quantity of water which it may hold or transmit.

If the strata consist of sand, water will pass through them

with facility, and they will also hold a considerable quantity between the interstices of their component grains ; whereas a bed of pure clay will not allow of the passage of water.   These are the two extremes of the case; the intermixture of these materials in the same bed will of course, according to the relative proportions, modify the transmission of water.   Prestwich found by experiment that a silicious sand of ordinary character will hold on an average rather more than one-third of its bulk of water, or from two to two and a half gallons in one cubic foot.   In strata so composed the water may be termed free, as it passes easily in all directions, and under the pressure of a column of water is comparatively but little impeded by capillary attraction.   These are the conditions of a true permeable stratum.   Where the strata are more compact and solid, as in sandstone, limestone, and oolite, although all such rocks imbibe more or less water, yet the water so absorbed does not pass freely through the mass, but is held in the pores of the rock by capillary attraction, and parted with very slowly; so that in such deposits water can be freely transmitted only in the planes of bedding and in fissures.   If the water-bearing deposit is of uniform lithological character over a large area, then the proposition is reduced to its simplest form; but when, as in the deposit between the London clay and chalk, the strata consist of variable mineral ingredients, it becomes essential to estimate the extent of these variations; for very different conclusions might be drawn from an inspection of the Lower Tertiary strata at different localities.

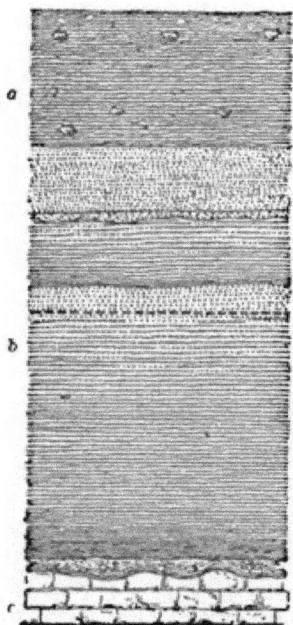

a London clay.   b Sands and clay.   c Chalk.
Fig. 5.

In the fine section exposed in the cliffs between Herne Bay and the Reculvers, in England, a considerable mass of fossiliferous sands is seen to rise from beneath the London clay. Fig. 5 represents a view of a portion of this cliff a mile and a half east of Herne Bay and continued downwards, by estimation, below the surface of the ground to the chalk. In this section there is evidently a very large proportion of sand, and consequently a large capacity for water. Again, at Upnor, near Rochester, the sands marked 3 are as much as 60 to 80 feet thick, and continue so to Gravesend, Purfleet, and Erith. In the first of these places they may be seen capping Windmill Hill; in the second, forming the hill, now removed, on which the lighthouse is built; and in the third, in the large ballast pits on the banks of the river Thames. The average thickness of these sands in this district may be about 50 to 60 feet. In their range from east to west, the beds 2 become more clayey and less permeable, and 1, very thin. As we approach London the thickness of 3 also

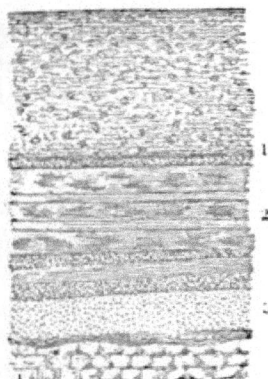

Fig. 6.

diminishes. In the ballast pits at the west end of Woolwich, this sand bed is not more than 35 feet thick, and as it passes under London becomes still thinner.

Fig. 6 is a general or average section of the strata on which London stands. The increase in the proportion of the argillaceous strata, and the decrease of the beds of sand in the Lower Tertiary strata, is here very apparent, and from this point westward to Hungerford, clays decidedly predominate; while at the same time the series presents such rapid variations, even on the same level and at short distances, that no two sections are alike. On the southern boundary of the

Fig. 7.

Tertiary district, from Croydon to Leatherhead, the sands 3 maintain a thickness of 20 to 40 feet, whilst the associated beds of clay are of inferior importance. We will take another section, Fig. 7, representing the usual features of the deposit in the northern part of the Tertiary district. It is from a cutting at a brickfield west of the small village of Hedgerley, 6 miles northward of Windsor.

Here we see a large development of the mottled clays, and but little sand. A somewhat similar section is exhibited at Oak End, near Chalfont St. Giles. But to show how rapidly this series changes its character, the section of a pit only a third of a mile westward of the one at Hedgerley is given in Fig. 8.

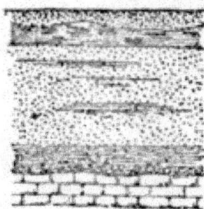

In this latter section the mottled clays have nearly disappeared, and are replaced by beds of sand with thin seams of mottled clays. At Twyford, near Reading, and at Old Basing, near Basingstoke, the mottled clays again occupy, as at Hedgerley, nearly the whole space between the London clays and the chalk. Near Reading a good section of these beds was exhibited in the Sonning cutting of the Great Western Railway; they consisted chiefly of mottled clays. At the Kats-grove pits, Reading, the beds are more sandy. Referring back to Fig. 6, it may be noticed that there is generally a small quantity of water found in the bed marked 1, in parts of the neighbourhood of London. Owing, however, to the constant presence of green and ferruginous sands, traces of vegetable matters and remains of fossil shells, the water is usually indifferent and chalybeate. The well-diggers term this a slow spring. They graphically express the difference by saying that the water creeps up from this stratum, whereas that it bursts up from the lower sands 3, which form the great water-bearing stratum. In the irregular sand beds interstratified with the mottled clays between these two strata water is also found, but not in any large quantity.

Fig. 9 is a section at the western extremity of the Tertiary district at Pebble Hill, near Hungerford. Here again the mottled clays are in considerable force, sands forming the smaller part of the series.

Fig. 8.

The following lists exhibit the aggregate thickness of all the beds of sand occurring between the London clay and the chalk, at various localities in the Tertiary district. It will appear from them that the mean results of the whole is very different from any of those obtained in separate divisions of the country. The mean thickness of the deposit throughout the whole Tertiary area may be taken at 62 feet, of which 36 feet consist of sands and 26 feet of clays; but as only a portion of this district contributes

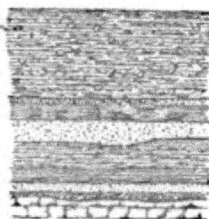

Fig. 9.

to the water supply of London, it will facilitate our inquiry if we divide it into two parts, the one westward of and including London, and the other eastward of it, introducing also some further subdivisions into each.

MEASUREMENT OF SECTIONS EASTWARD OF LONDON.

Southern Boundary.	Sand.	Clay.	Northern Boundary.	Sand.	Clay.
	ft.	ft.		ft.	ft.
Lewisham	65	26	Hertford	26	3
Woolwich	66	18	Beaumont Green, near⎫	16	10
Upnor	80?	8	Hoddesdon.. .. ..⎭		
Herne Bay	70?	50	Broxbourne	28	2
			Gestingthorpe, near Sud-⎫	50?	?
			bury .. ..⎭		
			Whitton, near Ipswich ..	60?	5
Average	70	25	Average	36	5

The mean of the three columns in two western sections gives a thickness to this formation of 57 feet, of which only 19 feet are sand and permeable to water, and the remaining 38 feet consist of impermeable clays, affording no supply of water.

The area, both at the surface and underground, over which they extend is about 1086 square miles.

The average total thickness of the eastern district deduced from the nine sections we have taken gives 68 feet, of which 53 feet are sands and 15 feet clays. The larger area, 1849

MEASUREMENT OF SECTIONS WESTWARD OF LONDON.

On or near the Southern Boundary of the Tertiary District.

	Sand. ft.	Clay. ft.
Streatham	30	25
Mitcham	47	34
Croydon	35?	20?
Epsom	31	23
Fetcham	35	20
Guildford	10	40
Chinham, near Basingstoke	20?	30
Itchingswell, near Kingsclere ..	22	34
Higheclere	24	27
Pebble Hill, near Hungerford	9	39
Average	26	29

On a Central Line in the Tertiary District.

	Sand. ft.	Clay. ft.	Sand. ft.	Clay. ft.
London:				
Millbank	49	40		
Trafalgar Square	49	30		
Tottenham Court Road ..	35	30		
Pentonville	34	44	46	30
Barclay's Brewery	55	42		
Lombard Street ..	53	35		
The Mint	49	38		
Whitechapel ..	45	50		
Garrett, near Wandsworth ..			20	52
Isleworth			17	70
Twickenham			7	50
Chobham			3	15
Average ..			18	51

On or near the Northern Boundary of the Tertiary District.

	Sand. ft.	Clay. ft.
Hatfield	23	2
Watford	25	10
Pinner	12	32
Oak End, Chalfont St. Giles	3	40
Hedgerley, near Slough	5	45
Starveall, ,, ,,	13	20
Twyford	5	60
Sonning, near Reading	12	54
Reading	16	33
Newbury	20	36
Pebble Hill	9	39
Average ..	13	34

square miles, over which the eastern portion of the Tertiary series extends, and the greater volume of the water-bearing beds, constitute important differences in favour of this district; and if there had been no geological disturbances to interfere with the continuous character of the strata, we might have looked to this quarter for a large supply of water to the Artesian wells of London.

From these tables it will be readily perceived that the strata of which the water-bearing deposits are composed are very variable in their relative thickness. They consist, in fact, of alternating beds of clay and sand, in proportions constantly changing. In one place, as at Hedgerley, the aggregate beds of sand may be 5 feet thick, and the clays 45 feet; whilst at another, as at Leatherhead, the sands may be 35, and the clays 20 feet thick, and some such variation is observable in every locality. But although we may thus in some measure judge of the capacity of these beds for water, this method fails to show whether the communication from one part of the area to another is free, or impeded by causes connected with mineral character. Now as we know that these beds not only vary in their thickness, but that they also frequently thin out, and sometimes pass one into another, it may happen that a very large development of clay at any one place may altogether stop the transit of the water in that locality. Thus in Fig. 10 the beds of sand at y allow of the free passage of water, but at x, where clays occupy the whole thickness, it cannot pass; the obstruction which this cause may offer to the underground flow of water can only be determined by experience. It must not, however, be supposed that such a variation in the strata is permanent or general along any given line. It is always local, some of the beds of clay commonly

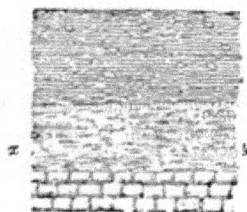

Fig. 10.

thinning out after a certain horizontal range, so that, although the water may be impeded or retarded in a direct course, it most probably can, in part or altogether, pass round by some point where the strata have not undergone the same alteration.

c

POSITION AND GENERAL CONDITIONS OF THE OUTCROP.

This involves some considerations to which an exact value cannot at present be given, yet which require notice, as they to a great extent determine the proportion of water which can pass from the surface into the mass of the water-bearing strata. In the first place, when the outcrop of these strata occurs in a valley, as represented in Fig. 11, it is evident that *b* may not

Fig. 11.

only retain all the water which might fall on its surface, but also would receive a proportion of that draining off from the strata of *a* and *c*. This form of the surface generally prevails wherever the water-bearing strata are softer and less coherent than the strata above and below them.

It may be observed in the Lower Tertiary series at Sutton, Carshalton, and Croydon, where a small and shallow valley, excavated in these sands and mottled clays, ranges parallel with the chalk hills.

It is apparent again between Epsom and Leatherhead, and also in some places between Guildford and Farnham, as well as between Odiham and Kingsclere. The Southampton railway crosses this small valley on an embankment at Old Basing.

This may be considered as the prevailing, but not exclusive, form of structure from Croydon to near Hungerford. The advantage, however, to be gained from it in point of water-supply is much limited by the rather high angle at which the strata are inclined, as well as by their small development, which greatly restrict the breadth of the surface occupied by the outcrop. It rarely exceeds a quarter of a mile, and is generally very much less, often not more than 100 to 200 feet.

The next modification of outcrop, represented in Fig. 12, is one not uncommon on the south side of the Tertiary district. The strata *b* here crop out on the slope of the chalk hills, and the rain falling upon them, unless rapidly absorbed, tends to drain

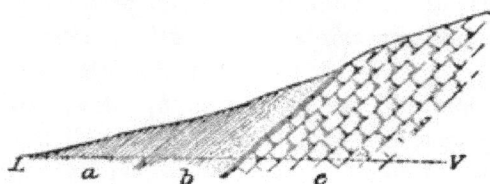

Fig. 12.

at once from their surface into the adjacent valleys. V, L, shows the line of valley level.

This arrangement is not unfrequent between Kingsclere and Inkpen, and also between Guildford and Leatherhead. Eastward of London it is exhibited on a larger scale at the base of the chalk hills in places between Chatham and Faversham, a line along which the sands of the Lower Tertiary strata, *b*, are more fully developed than elsewhere. As, however, the surface of *b* is there usually more coincident with the level, V, L, of the district, it is in a better position for retaining more of the rainfall.

A third position of outcrop, much more unfavourable for the water-bearing strata, prevails generally along the greater part of the northern boundary of the Tertiary strata. Instead of forming a valley, or outcropping at the base of the chalk hills, almost the whole length of this outcrop lies on the slope of the

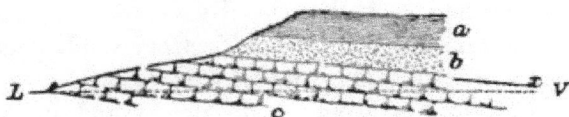

Fig. 13.

hills, as in Fig. 13, where the chalk, *c*, forms the base of the hill and the lower ground at its foot, while the London clay, *a*, caps

c 2

the summit, thus restricting the outcrop of b to a very narrow zone and a sloping surface. This form of structure is exhibited in the hills round Sonning, Reading, Hedgerley, Rickmansworth, and Watford; thence by Shenley Hill, Hatfield, Hertford, Sudbury; and also at Hadleigh this position of outcrop is continued. If, as on the southern side of the Tertiary district, the outcrop were continued in a nearly unbroken line, then these unfavourable conditions would prevail uninterruptedly; but the hills are in broken groups, and intersected at short distances by transverse valleys, as that of the Kennet at Reading, of the Loddon at Twyford, of the Colne at Uxbridge, and so on. Between Watford and Hatfield there is a constant succession of small valleys running back for short distances from the Lower district of the chalk, through the hills of the Tertiary district. The Valley of the Lea at Roydon and Hoddesdon is a similar and stronger case in point. The effect of these transverse valleys is to open out a larger surface of the strata b than would otherwise be exposed, for if the horizontal line, V, L, Fig. 13, were carried back beyond the point x, to meet the prolongation of b, then these Lower Tertiary strata would not only be intersected by the line of valley level, but would form a much smaller angle with the plain V, L, and therefore spread over a larger area than where they crop out on the side of the hills.

The foregoing are the three most general forms of outcrop, but occasionally the outcrop takes place wholly or partly on the summit of a hill, as, near the Reculvers in the neighbourhood of Canterbury, of Sittingbourne, and at the Addington Hills, near Croydon, in which cases the area of the Lower Tertiary is expanded. When the dip is very slight, and the beds nearly horizontal, the Lower Tertiary sands occasionally spread over a still larger extent of surface, as between Stoke Pogis, Burnham Common, and Beaconsfield, and in the case of the flat-topped hill, forming Blackheath and Bexley Heath, as in Fig. 14. Favourable as such districts might at first appear to be from the extent of their exposed surface, nevertheless they rarely contribute to the water supply of the wells sunk into the Lower Tertiary sands under London, the continuity of the strata being broken by intersecting valleys ; thus the district last mentioned is bounded on the north by the Valley of

Fig. 14.

the Thames, on the west by that of Ravensbourne, and on the east by the Valley of the Cray; consequently the rain-water, which has been absorbed by the very permeable strata on the intermediate higher ground, passes out on the sides of the hills, into the surface channels in the valleys, or into the chalk. Almost all the wells at Bexley Heath, for their supply of water, have, in fact, to be sunk into the chalk through the overlying 100 to 133 feet of sand and pebble beds, b.

Thus far we have considered this question, as if, in each instance, the outcropping edges of the water-bearing strata, b, were laid bare, and presented no impediment to the absorption of the rain-water falling immediately upon their surface, or passing on to it from some more impermeable deposits. But there is another consideration which influences materially the extent of the water supply.

If the strata b were always bare, we should have to consider their outcrop as an absorbent surface, of power varying according to the lithological character and dip of the strata only. But the outcropping edges of the strata do not commonly present bare and denuded surfaces. Thus a large extent of the country round London is more or less covered by beds of drift, which protect the outcropping beds of b, and turn off a portion of the water falling upon them.

The drift differs considerably in its power of interference with the passage of the rain-water into the strata beneath. The ochreous sandy flint gravel, forming so generally the subsoil of London, admits of the passage of water. All the shallow surface springs, from 10 to 20 feet deep, are produced by water which has fallen on, and passed through, this gravel, *g*, Fig. 15, down to the top of the London clay, *a*, on the irregular surface of which it is held up.

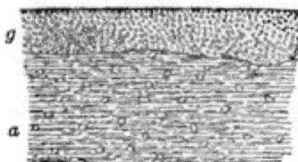

g, Gravel.   a, London clay.
Fig. 15.

When the London clay is wanting, this gravel lies immediately upon the Lower Tertiary strata, as in the valley between Windsor and Maidenhead, and in that of the Kennet between Newbury and Thatcham, transmitting to the underlying strata part of the surface water.

Where beds of brick earth occur in the drift, as between West Drayton and Uxbridge, the passage of the surface water into the underlying strata is intercepted.

Sometimes the drift is composed of gravel mixed very irregularly with broken-up London clay, and although commonly not more than 3 to 8 feet thick, it is generally impermeable.

Over a considerable portion of Suffolk and part of Essex, a drift, composed of coarse and usually light-coloured sand with fine gravel, occurs. Water percolates through it with extreme facility, but it is generally covered by a thick mass of stiff tenacious bluish-grey clay, perfectly impervious. This clay drift, or boulder clay, caps, to a depth of from 10 to 50 feet or more, almost all the hills in the northern division of Essex, and a large portion of Suffolk and Norfolk. It so conceals the underlying strata that it is difficult to trace the course of the outcrop of the Lower Tertiary sands between Ware and Ipswich; and often, as in Fig. 16, notwithstanding the breadth,

Fig. 16.

apart from this cause of the outcrop of the Tertiary sands, *b*, and of the drift of sand and gravel, 2, they are both so covered by the boulder clay, 1, that the small surface exposed can be of comparatively little value.

There are also, in some valleys, river deposits of silt, mud, and gravel. These are, however, of little importance to the subject before us. Under ordinary conditions they are generally sufficiently impervious to prevent the water from passing through the beds beneath.

HEIGHT OF WATER-BEARING STRATA ABOVE SURFACE OF COUNTRY.

The height of the districts, wherein the water-bearing strata crop out, above that of the surface of the country in which the wells are placed, should be made the subject of careful consideration, as upon this point depends the level to which the water in Artesian wells may ascend.

Again, taking the London district as an example, Prestwich remarks that, as the country rises on both sides of the Thames to the edge of the chalk escarpments, and as the outcrop of the Lower Tertiary strata is intermediate between these escarpments and the Thames, it follows that the outcrop of these lower beds must, in all cases, be on a higher level than the Thames itself, where it flows through the centre of the Tertiary district. Its altitude is, of course, very variable, as shown in the following list of its approximate height above Trinity high water-mark at London. These heights are taken where the Tertiaries are at their lowest level in the several localities mentioned.

South of London.			North of London.		
Croydon	about 130	feet.	Thetford	about 200	feet.
Leatherhead	„ 90	„	Watford	„ 170	„
Guildford	„ 96	„	Slough	„ 60	„
Old Basing	„ 250	„	Reading	„ 120	„
Near Hungerford	„ 360	„	Newbury	„ 236	„

Eastward of London these strata crop out at a gradually decreasing level. In consequence, therefore, of the outcrop of the water-bearing strata being thus much above the surface of the central Tertiary district bordering the Thames, the water in these strata beneath London tended originally to rise above that surface.

As, however, these beds crop out on a level with the Thames immediately east of the city between Deptford, Blackwall, and Bow, the water, having this natural issue so near, could never have risen in London much above the level of the river.

RAINFALL IN THE DISTRICT WHERE THE WATER-BEARING STRATA CROP OUT.

When inquiring into the probable relative value of any water-bearing strata, it is necessary to compare the rainfall in their respective districts.

Rain is of all meteorological phenomena the most capricious, both as regards its frequency and the amount which falls in a given time. In some places it rarely or never falls, whilst in others it rains almost every day; and there does not yet exist any theory from which a probable estimate of the rainfall in a given district can be deduced, independently of direct observation. But although dealing with one of the most capricious of the elements, we nevertheless find a workable average in the quantity of rain to be expected in any particular place, if careful and continued observations are made with the rain-gauge. G. J. Symons, the meteorologist, to whose continued investigations we are indebted for our most reliable data upon the subject of rainfall, gives the following practical instructions for using a rain-gauge :—

" The mouth of the gauge must be set quite level, and so fixed that it will remain so; it should never be less than 6 inches above the ground, nor more than 1 foot except when a greater elevation is absolutely necessary to obtain a proper exposure.

" It must be set on a level piece of ground, at a distance from shrubs, trees, walls, and buildings, at the very least as many feet from their base as they are in height.

" If a thoroughly clear site cannot be obtained, shelter is most endurable from N.W., N., and E., less so from S., S.E., and W., and not at all from S.W. or N.E.

"Special prohibition must issue as to keeping all tall-growing flowers away from the gauges.

"In order to prevent rust, it will be desirable to give the japanned gauges a coat of paint every two or three years.

"The gauge should, if possible, be emptied daily at 9 A.M., and the amount entered against the previous day.

"When making an observation, care should be taken to hold the glass upright.

"It can hardly be necessary to give here a treatise on decimal arithmetic; suffice it therefore to say that rain-gauge glasses usually hold half an inch of rain $(0 \cdot 50)$ and that each $\frac{1}{100}$ $(0 \cdot 01)$ is marked; if the fall is less than half an inch, the number of hundredths is read off at once; if it is over half an inch, the glass must be filled up to the half inch $(0 \cdot 50)$, and the remainder (say $0 \cdot 22$) measured afterwards, the total $(0 \cdot 50 + 0 \cdot 22) = 0 \cdot 72$ being entered. If less than $\frac{1}{10}$ $(0 \cdot 10)$ has fallen, the cipher must always be prefixed; thus if the measure is full up to the seventh line, it must be entered as $0 \cdot 07$, that is, no inches, no tenths, and seven hundredths. For the sake of clearness it has been found necessary to lay down an invariable rule that there shall always be two figures to the right of the decimal point. If there be only one figure, as in the case of one-tenth of an inch, usually written $0 \cdot 1$, a cipher must be added, making it $0 \cdot 10$. Neglect of this rule causes much inconvenience.

"In snow three methods may be adopted—it is well to try them all. 1. Melt what is caught in the funnel, and measure that as rain. 2. Select a place where the snow has not drifted, invert the funnel, and turning it round, lift and melt what is enclosed. 3. Measure with a rule the average depth of snow, and take one-twelfth as the equivalent of water. Some observers use in snowy weather a cylinder of the same diameter as the rain-gauge, and of considerable depth. If the wind is at all rough, all the snow is blown out of a flat-funnelled rain-gauge."

A drainage area is almost always a district of country enclosed by a ridge or watershed line, continuous except at the place where the waters of the basin find an outlet. It may be, and generally is, divided by branch ridge-lines into a number of smaller basins, each drained by its own stream into the main stream. In order to measure the area of a catchment basin a plan of the country is required, which either shows the ridge-lines or gives data for finding their positions by means of detached levels, or of contour lines.

When a catchment basin is very extensive it is advisable to measure the smaller basins of which it consists, as the depths of rainfall in them may be different; and, sometimes, also, for the same reason, to divide those basins into portions at different distances from the mountain chains, where rain-clouds are chiefly formed.

The exceptional cases, in which the boundary of a drainage area is not a ridge-line on the surface of the country, are those in which the rain-water sinks into a porous stratum until its descent is stopped by an impervious stratum, and in which, con-sequently, one boundary at least of the drainage area depends on the figure of the impervious stratum, being, in fact, a ridge-line on the upper surface of that stratum, instead of on the ground, and very often marking the upper edge of the outcrop of that stratum. If the porous stratum is partly covered by a second impervious stratum, the nearest ridge-line on the latter stratum to the point where the porous stratum crops out will be another boundary of the drainage area. In order to determine a drainage area under these circumstances it is necessary to have a geological map and sections of the district.

The depth of rainfall in a given time varies to a great extent at different seasons, in different years, and in different places. The extreme limits of annual depth of rainfall in different parts of the world may be held to be respectively nothing and 150 inches. The average annual depth of rainfall in different parts of Britain ranges from 22 inches to 140 inches, and the least annual depth recorded in Britain is about 15 inches.

The rainfall in different parts of a given country is, in general, greatest in those districts which lie towards the quarter from which the prevailing winds blow; in Great Britain, for instance, the western districts have the most rain. Upon a given mountain ridge, however, the reverse is the case, the greatest rainfall taking place on that side which lies to leeward, as regards the prevailing winds. To the same cause may be ascribed the fact that the rainfall is greater in mountainous than in flat districts, and greater at points near high mountain summits than at points farther from them; and the difference due to elevation is often greater by far than that due to 100 miles geographical distance.

The most important data respecting the depth of rainfall in a given district, for practical purposes, are, the least annual rainfall; mean annual rainfall; greatest annual rainfall; distribution of the rainfall at different seasons, and especially, the longest continuous drought; greatest flood rainfall, or continuous fall of rain in a short period.

The available rainfall of a district is that part of the total rainfall which remains to be stored in reservoirs, or carried away by streams, after deducting the loss through evaporation, through permanent absorption by plants and by the ground, and other causes.

The proportion borne by the available to the total rainfall varies very much, being affected by the rapidity of the rainfall and the compactness or porosity of the soil, the steepness or flatness of the ground, the nature and quality of the vegetation upon it, the temperature and moisture of the air, which will affect the rate of evaporation, the existence of artificial drains, and other circumstances. The following are examples:

Ground.	Available Rainfall ÷ Total Rainfall.
Steep surfaces of granite, gneiss, and slate,	nearly 1
Moorland and hilly pasture	from ·8 to ·6
Flat cultivated country	from ·5 to ·6
Chalk	0

Deep-seated springs and wells give from ·3 to ·4 of the total rainfall. Stephenson found that for the chalk district round Watford the evaporation was about 34 per cent., the quantity carried off by streams 23·2 per cent., leaving 42·8 per cent., which sank below the surface to form springs. In formations less absorbent than the chalk it can be calculated roughly, that streams carry off one-third, that another third evaporates, and that the remaining third of the total rainfall sinks into the earth.

Such data as the above may be used in approximately estimating the probable available rainfall in a district; but a much more accurate and satisfactory method is to measure the actual discharge of the streams, and the quantity lost by evaporation, at the same time that the rain-gauge observations are made, and so to find the actual proportion of available to total rainfall.

The following Table gives the mean annual rainfall in various parts of the world:—

TABLE OF RAINFALL. Collected by G. J. Symons.

Country and Station.	Period of Observations.	Latitude.		Mean Annual Fall.
	years.	°	′	ins.
EUROPE.				
AUSTRIA—Cracow	5	50	4 N	33·1
Prague	47	50	5	15·1
Vienna	10	48	12	19·6
BELGIUM—Brussels	20	50	51	28·6
Ghent	13	51	4	30·6
Louvain	12	50	33	28·6
DENMARK—Copenhagen	12	55	41	22·3
FRANCE—Bayonne	10	43	29	56·2
Bordeaux	32	44	50	32·4
Brest	30	48	23	38·8
Dijon	20	47	14	31·1
Lyons	..	45	46	37·0
Marseilles	60	43	17	19·0
Montpelier	51	43	36	30·3
Nice	20	43	43	55·2
Paris	44	48	50	22·9
Pau	12	43	19	37·1
Rouen	10	49	27	33·7
Toulon	..	43	4	19·7
Toulouse	52	43	36	24·9
GREAT BRITAIN—				
England, London	40	51	31	24·0
„ Manchester	40	53	29	36·0
„ Exeter	40	50	44	33·0
„ Lincoln	40	53	15	20·0

TABLE OF RAINFALL—*continued.*

Country and Station.	Period of Observations.	Latitude.	Mean Annual Fall.
EUROPE—*continued.*	years.	° ′	ins.
Wales, Cardiff	40	51 28 N	43·0
„ Llandudno	40	53 19	30·0
Scotland, Edinburgh	40	55 57	24·0
„ Glasgow	40	55 52	39·0
„ Aberdeen	40	57 8	31·0
Ireland, Cork	40	51 54	40·0
„ Dublin	40	53 23	30·0
„ Galway	40	53 15	50·0
HOLLAND—Rotterdam	..	51 55	22·0
ICELAND—Reikiavik	5	64 8	28·0
IONIAN ISLES—Corfu	22	39 37	42·4
ITALY—Florence	8	43 46	35·9
Milan	68	45 29	38·0
Naples	8	40 52	39·3
Rome	40	41 53	30·9
Turin	4	45 5	38·6
Venice	19	45 25	34·1
MALTA	..	35 54	15·0
NORWAY—Bergen	10	60 24	84·8
Christiania	..	59 54	26·7
PORTUGAL—Coimbra (in Vale of Mondego)	2	40 13	224·0 ?
Lisbon	20	38 42	23·0
PRUSSIA—Berlin	6	52 30	23·6
Cologne	10	50 55	24·0
Hanover	3	52 24	22·4
Potsdam	10	52 24	20·3
RUSSIA—St. Petersburg	14	59 56	16·2
Archangel	1	64 32	14·5
Astrakhan	4	46 24	6·1
Finland, Uleaborg	..	65 0	13·5
SICILY—Palermo	24	38 8	22·8
SPAIN—Madrid	..	40 24	9·0
Oviedo	1	43 22	111.1
SWEDEN—Stockholm	8	59 20	19·7
SWITZERLAND—Geneva	72	46 12	31·8
Great St. Bernard	43	45 50	58·5
Lausanne	8	46 30	38·5
ASIA.			
CHINA—Canton	14	23 6	69·3
Macao	..	22 24	68·3
Pekin	7	39 54	26·9
INDIA—			
Ceylon, Colombo	..	6 56	91·7
„ Kandy	..	7 18	84·0

TABLE OF RAINFALL—*continued.*

Country and Station.	Period of Observations.	Latitude.	Mean Annual Fall.
	years.	° ′	ins.
ASIA—*continued.*			
Ceylon, Adam's Peak ..	..	6 50 N	100·0
Bombay ..	33	18 56	84·7
Calcutta ..	20	22 35	66·9
Cherrapongee ..	..	25 16	610·3 ?
Darjeeling ..	..	27 3	127·3
Madras ..	22	13 4	44·6
Mahabuleshwur ..	15	17 56	254·0
Malabar, Tellicherry ..	..	11 44	116·0
Palamcotta ..	5	8 30	21·1
Patna ..	..	25 40	36·7
Poonah ..	4	18 30	23·4
MALAY—Pulo Penang ..	..	5 25	100·5
Singapore ..	..	1 17	190·0
PERSIA—Lencoran ..	3	38 44	42·8
Ooroomiah ..	1	37 28	21·5
RUSSIA—Barnaoul ..	15	53 20	11·8
Nertchinsk ..	12	51 18	17·5
Okhotsk ..	2	59 13	35·2
Tiflis ..	6	41 42	19·3
Tobolsk ..	2	58 12	23·0
TURKEY—Palestine, Jerusalem ..	{14	31 47	65·0 ?
	3	31 47	16·3
Smyrna ..	..	38 26	27·6
AFRICA.			
ABYSSINIA—Gondar ..	..	12 36	37·3
ALGERIA—Algiers ..	10	36 47	37·0
Constantina ..	..	36 24	30·8
Mostaganem ..	1	35 50	22·0
Oran ..	2	35 50	22·1
ASCENSION ..	2	8 8 S	11·5
CAPE COLONY—Cape Town ..	20	33 52	24·3
GUINEA—Christiansborg ..	..	5 30 N	19·2
MADEIRA ..	4	33 30	30·9
MAURITIUS—Port Louis ..	..	20 3 S	35·2
NATAL—Maritzburgh ..	..	29 36	27·6
ST. HELENA ..	3	15 55 N	18·8
SIERRA LEONE ..	..	8 30	86·0
TENERIFFE ..	2	28 28	22·3
NORTH AMERICA.			
BRITISH COLUMBIA—New Westminster ..	3	49 12	54·1
CANADA—Montreal, St. Martin's ..	2	45 31	47·3
Toronto ..	16	43 39	31·4

TABLE OF RAINFALL—*continued.*

Country and Station.	Period of Observations.	Latitude.	Mean Annual Fall.
NORTH AMERICA—*continued.*	years.	° ′	ins.
HONDURAS—Belize	1	17 29 N	153·0
MEXICO—Vera Cruz	..	19 12	66·1
RUSSIAN AMERICA—Sitka	7	57 3	89·9
UNITED STATES—Arkansas, Fort Smith ..	15	35 23	42·1
California, San Francisco	9	37 48	23·4
Nebraska, Fort Kearney	6	40 38	28·8
New Mexico, Socorro	2	34 10	7·9
New York, West Point	12	41 23	46·5
Ohio, Cincinnati	20	39 6	46·9
Pennsylvania, Philadelphia	19	39 57	43·6
South Carolina, Charleston	15	32 46	48·3
Texas, Matamoras	6	25 54	35·2
WEST INDIES—Antigua	..	17 3	39·5
Barbadoes	10	13 12	75·0
„ St. Philip	20	13 13	56·1
Cuba, Havannah	2	23 9	50·2
Matanzas	1	23 2	55·3
Grenada	..	12 8	126·0
Guadaloup, Basseterre	..	16 5	126·9
„ Matonba	..	16 5	285·8
Jamaica, Caraib	..	18 3	97·0
„ Kingstown	..	17 58	83·0
St. Domingo, Cape Haitien	..	19 43	127·9
„ Tivoli	..	19 0	106·7
Trinidad	..	10 40	62·9
Virgin Isles, St. Thomas'	..	18 17	60·6
„ Tortola	..	18 27	65·1
SOUTH AMERICA.			
BRAZIL—Rio Janeiro	..	22 54 S	58·7
S. Luis de Maranhao	..	3 0	276·0
GUYANA—Cayenne	6	4 56	138·3
Demerara, George Town	5	6 50	87·9
Parimaribo	..	6 0	229·2
NEW GRANADA—La Baja	6	7 22	54·1
Marmato	15	5 29	90·0
San Fé de Bogota	6	4 36	43·8
VENEZUELA—Cumana	..	10·27	7·5
Curaçoa	..	12·15 N	26·6
AUSTRALIA.			
NEW SOUTH WALES, Bathurst	3	33 24 S	22·7
Deniliquin	2	35 32	13·8

TABLE OF RAINFALL—*continued.*

Country and Station.	Period of Observations.	Latitude.	Mean Annual Fall.
	years.	° ′	ins.
AUSTRALIA—*continued.*			
NEW SOUTH WALES—Newcastle	3	32 57 S	55·3
Port Macquarie	12	31 29	70·8
Sydney	6	33 52	46·2
NEW ZEALAND—Auckland	2	36 50	31·2
Christchurch	3	43 45	31·7
Nelson	2	41 18	38·4
Taranaki	2	39 3	52·7
Wellington	2	41 17	37·8
SOUTH AUSTRALIA—Adelaide	6	34 55	19·2
TASMANIA—Hobart Town	12	42 54	20·3
VICTORIA—Melbourne	6	37 49	30·9
Port Phillip	11	38 30	29·2
WEST AUSTRALIA—Albany	..	35 0	32·1
York	1	31 55	25·4
POLYNESIA.			
SOCIETY ISLANDS—Tahiti Papiete	5	17 32	45·7

DISTURBANCES OF THE STRATA.

The last question to be considered relates to the disturbances which may have affected the strata; for whatever may be the absorbent power of the strata, the yield of water will be more or less diminished whenever the channels of communication have suffered break or fracture.

If the strata remained continuous and unbroken, we should merely have to ascertain the dimensions and lithological character in order to determine their actual water value. But if the strata is broken, the interference with the subterranean transmission of water will be proportionate to the extent of the disturbance.

Although the Tertiary formations around London have, probably, suffered less from the action of disturbing forces than the strata of any other district of the same extent in England, yet they nevertheless now exhibit considerable alterations from their original position.

The principal change has been that which, by elevation of the sides or depression of the centre of the district, gave the Tertiary deposits their present trough-shaped form, assuming it not to be the result of original deposition. If no further change had taken place we might have expected to find an uninterrupted communication in the Lower Tertiary strata from their northern outcrop at Hertford to their southern outcrop at Croydon, as well as from Newbury on the west to the sea on the east; and the entire length of 260 miles of outcrop would have contributed to the general supply of water at the centre.

But this is far from being the case; several disturbing causes have deranged the regularity of original structure. The principal one has caused a low axis of elevation, or rather a line of flexure running east and west, following nearly the course of the Thames from the Nore to Deptford, and apparently continued thence beyond Windsor. It brings up the chalk at Cliff, Purfleet, Woolwich, and Loampit Hill to varied but moderate elevations above the river level. Between Lewisham and Deptford the chalk disappears below the Tertiary series, and does not come to the surface till we reach the neighbourhood of Windsor and Maidenhead.

There is also, probably, another line of disturbance running between some points north and south and intersecting the first line at Deptford. It passes apparently near Beckenham and Lewisham, and then, crossing the Thames near Deptford, continues up a part, if not along the whole length of the valley of the Lea towards Hoddesdon. This disturbance appears in some places to have resulted in a fracture or fault in the strata, placing the beds on the east of it on a higher level than those on the west; and at other places merely to have produced a curvature in the strata. Prestwich states that he was unable to give its exact course, but its effect, at all events upon the water supply of London, is important, as, in conjunction with the first or Thames valley disturbance, it cuts off the supplies from the whole of Kent, and interferes most materially with the supply from Essex; for in its course up the valley of the Lea it either brings up the Lower Tertiary strata to the surface, as at Strat-

D

ford and Bow, or else, as farther up the valley, it raises them
to within 40 or 60 feet of the surface.

The Tertiary district thus appears, on a general view, to be
divided naturally into four portions by lines running nearly
north and south, the former line passing immediately south, and
the latter east of London, which stands at the south-east corner
of the north-western division, and consequently it must not be
viewed as the centre of one large and unbroken area, so far as
the Tertiary strata are concerned.

CHAPTER II.

JURASSIC STRATA.

UNDER the term Jurassic, derived from the Jura Mountains in Switzerland, are now grouped the great series of fossiliferous rocks which were formerly termed Oolitic, from the characteristic oolitic structure of many of its limestones. It includes all the beds between the Hastings sands and the New red sandstone. The strata of the Jurassic period in England appear at the surface over a narrow range of country, averaging 30 miles in width, commencing at Lyme Regis and Portland on the English Channel, and extending across England, north and north-east to the River Humber, and still further north, on the eastern coast of Yorkshire, almost to the mouth of the Tees. They thus cover eastern England.

The oolitic rocks are very porous, absorbing and holding enormous volumes of water, which are again delivered as springs, usually of great size. As water-bearing rocks the oolites are equal, if not superior, to the chalk itself for the purification and storage of water, but it is much to be regretted that this vast store is rarely used by communities in England until it has been hopelessly polluted. Analyses taken from time to time over the district show that in opening a well great care should be exercised to cut off surface communication in deep wells, and that most shallow wells are unsafe.

An area of not less than 6671 square miles is occupied by the oolitic rocks of England, with an annual average absorption of not less than 10 inches of rainfall, a figure probably much below the real average.

The two chief sources of springs among the Cotteswolds are the base of the Great oolite or Stonesfield slate, at its junction with the Fuller's earth, and at the junction of the Upper Lias

clay with the overlying sands. To the latter horizon belong
the seven springs forming the source of the Thames. Smaller
springs issue in the district at the base of the Lias Marlstones,
and the upper surfaces of the Forest Marble clays.

Gloucester is partially supplied by springs in the flanks of
Robin's Wood Hill thrown out by the lias, which, with the
surface drainage, are collected in a reservoir.

Three springs at Cheltenham are collected along the flanks
of the hills in bricked wells, and conveyed to the reservoirs at
Hewlett's Hill and Leckhampton, together holding 35,000,000
gallons.

Above 300,000 gallons are delivered daily, the water being
much softer than most oolitic springs, the hardness being only
15°·0 of which 6·0 is permanent; that of Haydon, near Cheltenham, is no less than 45°·7, of which 13·4 is permanent.

From springs off the Upper Lias, Bath is supplied with
water by no less than eighteen private companies. The water
derived from the Beacon springs is the best, but is not quite
satisfactory, since it is a remarkable fact that the cold waters, as
well as the thermal springs of Bath, have considerable organic
impurity.

The Great oolite at Bath, consists of the following series :—

		Feet.
Upper Rags .. { Coarse shelly limestone .. .. Fine grained oolite .. .. .. Tough brown limestone		20 to 55
Fine Freestone		10 to 30
Lower Rags .. Coarse shelly limestone		10 to 40

The freestone is very soft when first obtained, containing
much moisture, amounting sometimes, it is said, to one gallon
of water a cubic foot.

The Bradford clay, a local thickening of the clayey beds of
the overlying Forest Marble, reaches its greatest thickness at
Farleigh, where it is 40 to 60 feet. The Forest Marble around
Bath is 100 feet thick, in the Cotteswolds not more than 50.
The Cornbrash limestones reach a thickness of more than 40
feet, and are overlaid by 300 to 400 feet of Oxford clay.

The Coral rag, a rubbly limestone composed mainly of masses of coral, only appears in the Bristol area. At Longleat Park it underlies the Kimmeridge, which reaches a thickness of 65 feet at Maiden Bradley.

In the oolitic outcrop, ranging between Crewkerne, through Bath to Wotton-under-Edge, the Coral rag is water-bearing. When present, the Oxford clay forms the impermeable layer, as also does the Cornbrash and the upper sandy beds of the Forest Marble, which are held up by the clayey bed beneath.

The Lincolnshire oolites are absent in the eastern and southern portions of the Midland district, and the base of the Great oolite rests directly upon an eroded and denuded surface of the Northampton sands.

The basement beds of the Northampton sands rest in Rutland and South Lincoln on an eroded surface of Upper Lias clay, and generally consist of oolitic ironstone rock, forming a bold escarpment called "The Cliff," which stretches for 90 miles through Lincolnshire to Yorkshire. From its base, at the junction of the Lias, copious springs arise.

A boring at Stamford reached a depth of 500 feet, but the Lias was not penetrated, the upper clay being above 150 feet thick. Water occurs in the same horizon in the Uppingham Outlier, issuing from a blue calcareous rock, forming the base of the Northampton sands, at Lyddington. Springs also issue at Bisbrook.

The upper portion of the ironstones are much peroxidised and readily pervious to water; the compact lower portion, carbonate of iron, is the water-bearing horizon, but it is considered locally much safer to penetrate it and reach the Lias " blue bind " to prevent failure during droughts.

The Northampton sands average 20 to 30 feet in thickness, and seldom reach more than 40 feet. The overlying Lincolnshire oolite at Stamford is 80 feet, thickening from thence northwards, and thinning out entirely southwards at Harrington and Maidwell, and eastwards near Wansford tunnel.

At Northampton a recent bore-hole put down by the Water Company, at the Kettering road, commenced in the Lias clays, which extended to a depth of 738 feet, but below the Lias, instead of the triassic beds, a series of sandstones, conglomerates, and marls, terminating in carboniferous limestone, were met with.

As regards the quality of water derived from the oolitic rocks, selected analyses made for the Rivers Pollution Commission indicate that these rocks are not inferior to the New red sandstone, in the energy with which they oxidise and destroy the organic matter present in the waters percolating through them.

Though the waters so derived are generally hard, it is chiefly of a temporary character, capable of being softened by Clark's process, so as to average $6°·8$ instead of $20°·6$. The oolites yield, in springs and deep wells, water which is bright, sparkling, and palatable, excellent for drinking and all domestic purposes except washing, for which latter purpose the addition of lime renders it fit.

It is noticeable that the temporary hardness of the deep-well waters is higher than that of the spring water.

CHAPTER III.

THE NEW RED SANDSTONE.

THIS formation has been already alluded to at pp. 5 and 8; it is, next to the chalk and lower greensand, the most extensive source of water supply from wells we have in England, and although the two formations mentioned occupy a larger area, yet, owing to geographical position, the new red sandstone receives a more considerable quantity of rainfall, and, owing to the comparative scarceness of carbonate of lime, yields softer water.

The new red sandstone is called on the Continent "the Trias," as in Germany and parts of France it presents a distinct threefold division. Although the names of each of the divisions are commonly used, they are in themselves local and unessential, as the same exact relations between them do not occur in other remote parts of Europe or in England, and are not to be looked for in distant continents. The names of the divisions and their English equivalents are:

1. Keuper, or red marls.
2. Muschelkalk, or shell limestones (not found in this country).
3. Bunter sandstone, or variegated sandstone.

The strata consist in general of red, mottled, purple, or yellowish sandstones and marls, with beds of rock-salt, gypsum pebbles, and conglomerate.

The region over which triassic rocks outcrop in England stretches across the island from a point in the south-western part of the English Channel about Exmouth, Devon, north-north-eastward, and also from the centre of this band along a north-westward course to Liverpool, thence dividing and running north-east to the Tees, and north-west to Solway Firth.

In central Europe the trias is found largely developed, and

in North America it covers an area whose aggregate length is some 700 or 800 miles.

The beds, in England, may be divided as follows :

	Average Thickness.
KEUPER—Red marls, with rock-salt and gypsum	1000 ft.
Lower Keuper sandstones, with trias sandstones and marls (waterstones)	250 ft.
Dolomitic conglomerate	
BUNTER—Upper red and mottled sandstone	300 ft.
Pebble beds, or uncompacted conglomerate ..	300 ft.
Lower red and mottled sandstone	250 ft.

The Keuper series is introduced by a conglomerate often calcareous, passing up into brown, yellow, or white freestone, and then into thinly laminated sandstones and marls. The other subdivisions are remarkably uniform in character, except in the case of the pebble beds, which in the north-west form a light red pebbly building stone, but in the central counties become generally an unconsolidated conglomerate of quartzose pebbles.

The following tabulated form, due to Edward Hull, shows the comparative thickness and range of the Triassic series along a south-easterly direction from the estuary of the Mersey, and also shows the thinning away of all the Triassic strata from the north-west towards the south-east of England, which Hull was amongst the first to demonstrate.

THICKNESS AND RANGE OF THE TRIAS IN A S.E. DIRECTION FROM THE MERSEY.

Names of Strata.	Lancashire and West Cheshire.	Staffordshire.	Leicestershire and Warwickshire.
KEUPER SERIES—Red Marl	3000	800	700
Lower Keuper sandstone	450	200	150
BUNTER SERIES—Upper mottled sandstone	500	50 to 200	absent
Pebble beds	500 to 750	100 to 300	0 to 100
Lower mottled sandstone	200 to 500	0 to 100	absent

The formation may be looked upon as almost equally permeable in all directions, and the whole mass may be regarded as a reservoir up to a certain level, from which, whenever wells are sunk, water will always be obtained more or less abundantly. This view is very fairly borne out by experience, and the occurrence of the water is certainly not solely due to the presence of the fissures or joints traversing the rock, but to its permeability, which, however, varies in different districts.

In the neighbourhood of Liverpool the rock, or at least the pebble bed, is less porous than in the neighbourhood of Whitmore, Nottingham, and other parts of the midland counties, where it becomes either an unconsolidated conglomerate or a soft crumbly sandstone. Yet wells sunk even in the hard building stone of the pebble beds, either in Cheshire or Lancashire, always yield water at a certain variable depth. Beyond a certain depth the water tends to decrease, as was the case in the St. Helen's public well, situated on Eccleston Hill. At this well an attempt was made, in 1868, to increase the supply by boring deeper into the sandstone, but without any good result.

When water percolates downwards in the rock we may suppose there are two forces of an antagonistic character brought into play; there is the force of friction, increasing with the depth, and tending to hinder the downward progress of the water, while there is the hydrostatic pressure tending to force the water downwards; and we may suppose that when equilibrium has been established between these two forces, the further percolation will cease.

The proportion of rain which finds its way into the rock in some parts of the country must be very large. When the rock, as is generally the case in Lancashire, Cheshire, and Shropshire, is partly overspread by a coating of dense boulder clay, almost impervious to water, the quantity probably does not exceed one-third of the rainfall over a considerable area; but in some parts of the midland counties, where the rock is very open, and the covering of drift scanty or altogether absent, the percolation amounts to a much larger proportion, probably one-half or two-thirds, as all the rain which is not evaporated passes downwards. The new red sandstone, as remarked, may be regarded, in respect to water supply, as a nearly homogeneous mass, equally available throughout; and it is owing to this structure, and the almost entire absence of beds of impervious clay or marl, that the formation is capable of affording such large supplies of water; for the rain which falls on its surface and penetrates into the rock is free to pass in any direction towards a well when sunk in a central position. If we consider the rock as a mass completely saturated with water through a certain vertical depth, the water being in a state of equilibrium, when a well is sunk, and the water pumped up, the state of equilibrium is destroyed, and the water in the rock is forced in from all sides. The percolation is, doubtless, much facilitated by joints, fissures, and faults, and in cases where one side of a fault is composed of impervious strata, such as the Keuper marls, or coal measures, the quantity of water pent up against the face of the fault may be very large, and the position often favourable for a well.

An instance of the effect of faults in the rock itself, in increasing the supply, is afforded in the case of the well at Flaybrick Hill, near Birkenhead. From the bottom of this well a heading was driven at a depth of about 160 feet from the surface, to cut a fault about 150 feet distant, and upon this having been effected the water flowed in with such impetuosity that the supply, which had been 400,000 gallons a day, was at once doubled.

The water from the new red sandstone is clear, wholesome, and pleasant to drink; it is also well adapted for the purposes of bleaching, dyeing, and brewing; at the same time it must be admitted that its qualities as regards hardness, in other words, the proportions of carbonate of limes and magnesia it contains, are subject to considerable variation, depending on the locality and composition of the rock. As a general rule the water from the new red sandstone may be considered as occupying a position intermediate between the *hard* water of the chalk, and the *soft* water supplied to some of our large towns from the drainage of mountainous tracts of the primary formations, of which

the water supplied from Loch Katrine to Glasgow is perhaps
the purest example, containing only 2·35 grains of solid matter
to the gallon. Having besides but a small proportion of saline
ingredients, which, while they tend to harden the water, are
probably not without benefit in the animal economy, the water
supply from the new red sandstone possesses incalculable advantages over that from rivers and surface drainage. Many of
our large towns are now partially or entirely supplied with
water pumped from deep wells in this sandstone; and several
from copious springs gushing forth from the rock at its junction
with some underlying impervious stratum belonging to the
primary series.

CHAPTER IV.

WELL SINKING.

Previous to sinking it will be necessary to have in readiness a stock of buckets, shovels, picks, rope, a pulley-block or a windlass, and barrows or other means of conveying the material extracted away from the mouth of the sinking. If the sinking is of any great depth, a few lengths of portable railway and tipping waggons will be of much service. After all the preliminary arrangements have been made, the sinking is commenced by marking off a circle upon the ground 12 or 18 inches greater in circumference than the intended internal diameter of the well. The centre of the well as commenced from must be the centre of every part of the sinking; its position must be carefully preserved, and everything that is done must be true to this centre, the plumb-line being frequently used to test the vertical position of the sides.

To sink a well by underpinning, an excavation is first made to such a depth as the strata will allow without falling in. At the bottom of the excavation is laid a curb, that is, a flat ring, whose internal diameter is equal to the intended clear diameter of the well, and its breadth equal to the thickness of the brickwork. It is made of oak or elm planks 3 or 4 inches thick, either in one layer fished at the joints with iron, or in two layers breaking joint, and spiked or screwed together. On this, to line the first division of the well, a cylinder of brickwork, technically called steining, is built in mortar or cement. In the centre of the floor is dug a pit, at the bottom of which is laid a small platform of boards; then, by cutting notches in the side of the pit, several raking props are inserted, their lower ends abutting against a foot block, and their upper ends against the lowest setting, so as to give temporary support to the curb with its load of brickwork. The pit is enlarged to the diameter of the shaft above; on the bottom of the excavation is laid a new curb, on which is built a new division of the brickwork, giving permanent support to the upper curb; the raking props and their foot-blocks are removed; a new pit is dug, and so on as before. Care should be taken that the earth is firmly packed behind the steining.

A common modification of this method consists in excavating to such a depth as the strata will admit without falling in. A wooden curb is laid at the bottom of the excavation, the brick steining laid upon it and carried to the surface. The earth is then excavated flush with the interior sides of the well, so that the earth underneath the curb supports the brickwork above. When the excavation has been carried on as far as convenient, recesses are made in the earth under the previous steining, and in these recesses the steining is carried up to the previous work. When thus supported the intermediate portions of earth between the sections of brickwork carried up are cut away and the steining completed.

In sinking with a drum curb, the curb, which may be either of wood or iron, consists of a flat ring for supporting the steining, and of a vertical hollow cylinder or drum of the same outside diameter as the steining, supporting the ring within it and bevelled to a sharp edge below. The rings, or ribs, of a wooden curb are formed of two thicknesses of elm plank, $1\frac{1}{2}$ inch thick by 9 inches wide, giving a total thickness of 3 inches.

Fig. 17 is a plan of a wooden drum curb, and Fig. 18 a section showing the mode of construction. The outside cylinder or drum is termed the lagging, and is commonly made from $1\frac{1}{2}$-inch yellow pine planks. The drum may be strengthened if necessary by additional rings, and its connections with the rings made more secure by brackets. In large curbs the rings are placed about 3 feet 6 inches apart. Fig. 19 is a section of such a curb for a 20-feet opening. Here the rings are each

Fig. 17.

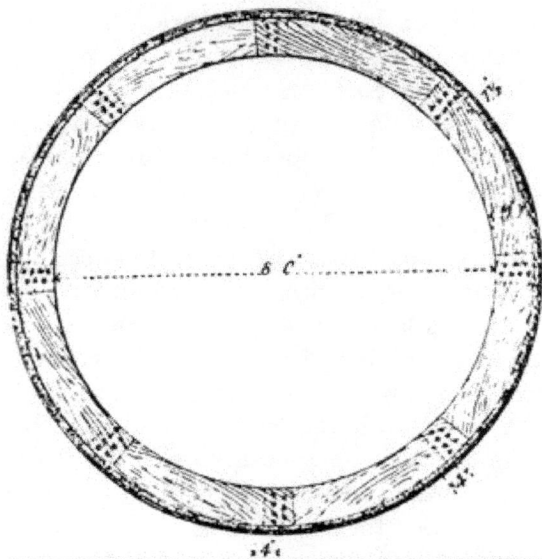

Fig. 18.

three deep, and of such thickness as to afford strength to resist
great lateral pressure. Fig. 20 is a plan, and Fig. 21 an
enlarged segment of an iron curb.

When the well has been sunk as far as the earth will stand
vertical, the drum curb is lowered into it and the building of
the brick cylinder com-
menced, care being taken
to complete each course
of bricks before laying
another, in order that
the curb may be loaded
equally all round. The
earth is dug away from
the interior of the drum,
and this, together with
the gradually increasing
load, causes the sharp
lower edge of the drum
to sink into the earth ;
and thus the digging of
the well at the bottom,
the sinking of the drum
curb and the brick lin-

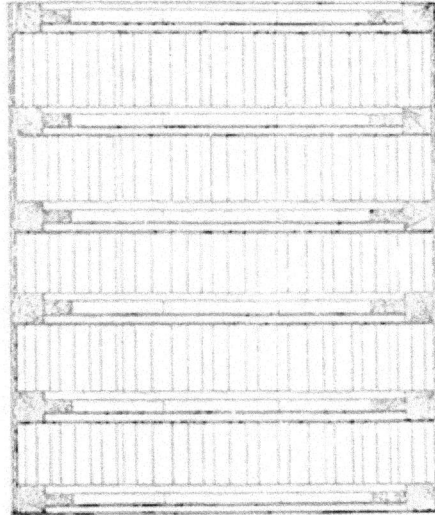

Scale ¹⁄₁₂ inch = 1 foot.
Fig. 19.

ing which it carries, and the building of the steining at the
top, go on together. Care must be taken in this, as in every
other method, to regulate the digging so that the well shall sink
vertically. Should the friction of the earth against the outside
of the well at length become so great as to stop its descent
before the requisite depth is attained, a smaller well may be
sunk in the interior of the first well. A well so stopped is said
to be earth-fast. This plan cannot be applied to deep wells,
but is very successful in sandy soils where the well is of
moderate depth.

The curbs are often supported by iron rods, fitted with screws
and nuts, from cross timbers over the mouth of the well, and as
the excavation is carried on below, brickwork is piled on above,
and the weight of the steining will carry it down as the excava-

tion proceeds, until the friction of the sides overpowers the
gravitating force, when it becomes earth-bound ; then a set-off
must be made in the well, and the same operation repeated as

Fig. 20.

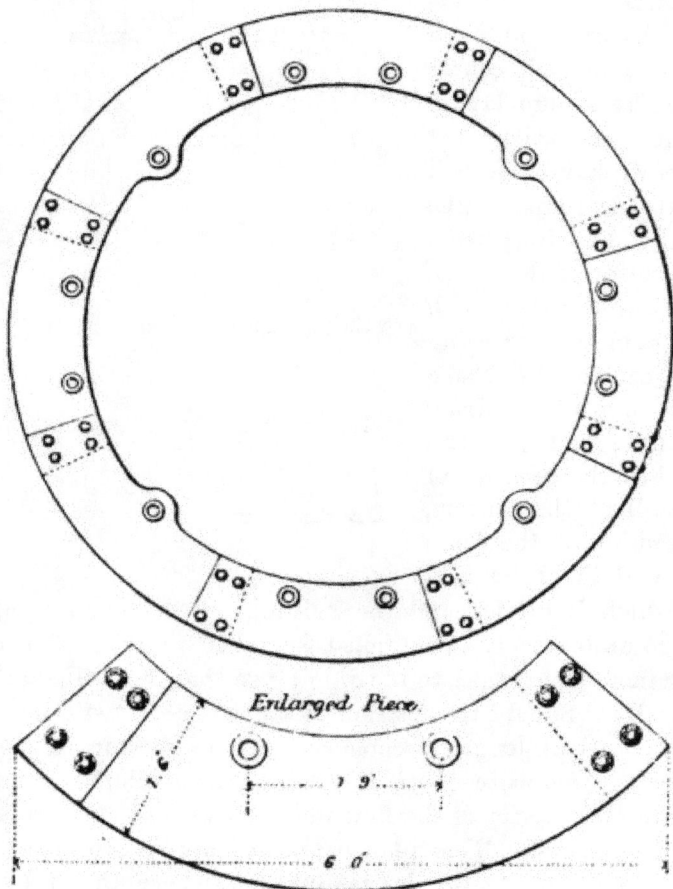

Enlarged Piece.

Fig. 21.

often as the steining becomes earth-bound, or the work must be
completed by the first method of underpinning.

When the rock to be sunk through is unstratified, or if stratified, when of great thickness, recourse must be had to the action of explosive agents. The explosives most frequently used for this purpose are gunpowder, guncotton, and dynamite. Of these gunpowder is the oldest and still one of the most extensively employed, and although the more violent explosives are so much used, it is not at all probable that gunpowder will ever be entirely displaced by them as a blasting material, no other explosive agent possesses its peculiar properties or can be used instead of it under all circumstances. It is essential, however, that the powder be of good quality, a matter which is much too frequently neglected.

The advantages, in certain cases, of a stronger explosive than gunpowder led to the introduction of the nitro-cotton and the nitro-glycerine preparations, and of these dynamite, the name given to nitro-glycerine absorbed in powdered kieselguhr or infusorial silica, is the most generally useful. In very hard and tough rock it is very effective, and will bring out a burden which other explosives fail to loosen. It is not much affected by damp, so that it may be employed in wet holes; indeed, water is commonly used as a tamping with this explosive. In upward holes, where water cannot of course be used, dynamite is fired without tamping, its quick action rendering this possible, although it is more economical to use light tamping.

The plastic form of dynamite constitutes a great practical advantage, inasmuch as it allows the explosive to be rammed tightly into the bore-hole, so as to fill up all empty spaces and crevices; this also renders it very safe to handle, as a light blow can hardly produce sufficient heat in it to cause an explosion.

The numerous other mixtures of nitro-glycerine, such as "mica-powder," "rend-rock," "litho-fracteur," and the like, may be considered as dynamites and employed in the same way. Blasting-gelatine, which consists of nitro-glycerine gelatinised by the addition of soluble guncotton, requires a very strong detonator, or a primer cartridge of another explosive to produce its best results.

E

Dynamite and guncotton have to be fired in a different way to gunpowder, since a spark or the mere application of flame will not cause them to explode. A detonator or powerful cap is the means employed, and this is attached either to the ordinary safety fuse, or to an electric fuse. The fuse fires the detonator, which explodes and fires the explosive.

When burnt unconfined, dynamite and guncotton give no practical effect, but evolve fumes that are very disagreeable; if properly detonated by using a detonator of sufficient strength, and placing it well into the cartridge, and if over-charging be avoided, their explosion will not vitiate the atmosphere. So-called "treble" detonators are best for dynamite, "quintuples" for guncotton, and "sextuples" for tonite or cotton-powder.

The following instructions for using dynamite are those usually given by the writer; they will apply almost equally to any nitro-glycerine mixture, or to guncotton and its derivative, cotton-powder. A piece of suitable safety-fuse is taken, a sufficient length cut off cleanly, and the end put into a deto-nator. The detonator must be attached firmly by squeezing it on to the fuse with a pair of nippers, at the end nearest the fuse. This is a matter of importance, since the squeezing not only retains the detonator in its place, but enables its full force to be utilised. In wet ground or water, a little tar, grease, or red lead should be smeared round the junction of the fuse and detonator, to prevent the admission of moisture into the latter, which might cause a missfire. Open a primer cartridge at one end, and with a small pointed piece of wood make a hole in the dynamite about $\frac{1}{2}$ inch deep; put the detonator into this hole, leaving the upper part of the cap quite clear of the explosive; then twist the paper round the fuse, Fig. 22, and tie it firmly with a piece of string, so that the detonator may not be pulled out when the primer is in the bore-hole. The primer should in every case be of a diameter inferior to that of the

Fig. 22

shot-hole by at least one-eighth or, better still, one-fourth part of an inch. The reason for this is that the primer to a shot should always be of such a size as will not necessitate the employment of any material degree of force in placing it in position on to the top of the charge, the possibility of exploding the detonator by means of a blow in the act of charging a shot is thereby avoided. Detonators, being simply percussion caps of large size and power, will explode as surely from the effects of a blow as from the application of fire. When primers of small diameter in comparison with the size of the shot-hole are made use of, the only slight source for fear is avoided.

The requisite number of plain cartridges having been taken, each is pressed separately into the bore-hole with a wooden rammer, Fig. 23. The quantity required will vary with the size and depth of the bore-hole, and the kind of rock to be blasted; a few shots will easily determine this in any particular locality. When all the charge is inserted, the primer, with the detonator and fuse, is gently pushed upon the top and lightly tamped with sand, clay, or even water. There should be a depth of not less than 2 inches of sand over the end of the detonator before

Fig. 23.

commencing to tamp. Though a high opinion is entertained by practical miners of the value of hard tamping in the use of gunpowder, it is a well-ascertained fact that when dynamite or guncotton is the explosive used, gentle tamping answers well, though as the depth of tamping in the shot-hole increases, the tamping may become heavier with advantage. Under any circumstances, wooden tamping-rods should alone be used.

Care must be taken that there is no foreign substance in the tube of the detonator, also to have the fuse cut off level at the end that is to be inserted into the detonator, thus ensuring that the small column of prepared gunpowder of the fuse shall come in direct contact with the fulminate.

Another point that requires attention is the operation of fixing and securing the detonator into the primer. Not only is it necessary that the detonator should be securely fixed, but when so fixed not less than, say, one-fourth of the total length of the detonator should be visible. Should the detonator become more than fully inserted into the primer, the fuse might possibly set on fire the primer before exploding the detonator, and thus cause a comparative failure of the shot.

Fig. 24.

Having made the arrangements thus described, the bore-hole will be similar to Fig. 24; it is then ready to fire.

In winter time, or whenever the temperature is low, dynamite freezes and becomes hard. It should not be used until it is thawed, when it softens and is again fit for use. The thawing may be easily done by putting the cartridges in a tin can, and this in an outer vessel containing hot water, or in the hot water cans supplied by the manufacturers. A rough but safe plan in careful hands, is to keep them a little time in the trousers pockets.

On no account should the dynamite be warmed on iron plates, stoves, or before an open fire, as such practices are a most fruitful cause of serious accidents. For the same reason open boxes of dynamite should not be exposed to the sun.

Dynamite for miners' use is commonly made up in charges of $\frac{7}{8}$ to 2 inches diameter, advancing by eighths. The ordinary size is the 1-inch.

A 1-inch bore-hole will contain 8·18 ounces of dynamite in every foot of its length, and for estimating purposes 18 coils of safety fuse and 200 detonators may be allowed to each 100 lb. of explosive.

We shall, in treating generally of blasting for well sinking, consider these operations as carried out by the aid of gunpowder. Similar reasoning will apply to the other explosives.

The system of blasting employed in well sinking consists in boring holes from $\frac{7}{8}$ to 3 inches diameter in the rock to be disrupted, to receive the charge. The position of these holes is a matter of the highest importance from the point of view of producing the greatest effects with the available means, and to determine them properly requires a complete knowledge of the nature of the forces developed by an explosive agent. This knowledge is rarely possessed by sinkers. Indeed, such is the ignorance of this subject displayed by quarrymen generally, that when the proportioning and placing the charges are left to their judgment, a large expenditure of labour and material will often produce very inadequate results. In all cases it is far more economical to entrust those duties to one who thoroughly understands the subject. The following principles should govern all operations of this nature :

The explosion of gunpowder, by the expansion of the gases suddenly evolved, develops an enormous force, and this force, due to the pressure of a fluid, is exerted equally in all directions. Consequently the surrounding mass subjected to this force will yield, if it yield at all, in its weakest part, that is, in the part which offers least resistance. The line along which the mass yields, or line of rupture, is called the line of least resistance, and is the distance traversed by the gases before reaching the surface. When the surrounding mass is uniformly resisting, the line of least resistance will be a straight line, and will be the shortest distance from the centre of the charge to the surface. Such, however, is rarely the case, and the line of rupture will therefore in most instances be an irregular line, and often much longer than that from the centre direct to the surface. Hence in all blasting operations there will be two things to determine, the line of least resistance and the quantity of powder requisite to overcome the resistance along that line. For it is obvious that all excess of powder is waste; and, moreover, as the force developed by this excess must be expended upon something, it will probably be employed in doing mischief. Charges of powder of uniform strength produce effects varying with their weight, that is, a double charge will move a double mass. And as homogeneous masses vary as the cube of any similar line within them, the general rule is established that charges of powder to produce similar results are to each other as the cube of the lines of least resistance.   Hence when the charge requisite to produce a given effect in a particular substance has been determined by experiment, that necessary to produce a like effect in a given mass of the same substance may be readily determined.   As the substances to be acted upon are various and differ in tenacity in different localities, and as, moreover, the quality of powder varies greatly, it will be necessary, in undertaking sinking operations, to make experiments in order to determine the constant which should be employed in calculating the charges of powder.   In practice, the line of least resistance is taken as the shortest distance from the centre of the charge to the surface of the rock, unless the existence of natural divisions shows it to lie in some other direction ; and, generally, the charge requisite to overcome the resistance will vary from $\frac{1}{15}$ to $\frac{1}{35}$ of the cube of the line, the latter being taken in feet, and the former in pounds.   Thus, suppose the material to be blasted is chalk, and the line of least resistance 4 feet, the cube of 4 is 64, and taking the proportion for chalk as $\frac{1}{30}$, we have $\frac{64}{30} = 2\frac{2}{13}$ lb. as the charge necessary to produce disruption.

When the blasting is in stratified rock, the position of the charge will frequently be determined by the natural divisions and fissures ; for if these are not duly taken into consideration, the sinker will have the mortification of finding, after his shot has been fired, that the elastic gases have found an easier vent through one of these flaws, and that consequently no useful effect has been produced. The line of least resistance, in this case, will generally be perpendicular to the beds of the strata, so that the hole for the charge may be driven parallel to the strata and in such a position as not to touch the planes which separate them. This hole should never be driven in the direction of the line of least resistance, and when practicable should be at right angles to it.

The instruments employed in boring the holes for the shot

are iron rods having a wedge-shaped piece of steel welded to their lower ends and brought to an edge so as to cut into the rock. These are worked either by striking them on the head with a hammer, or by jumping them up and down and allowing them to penetrate by their own weight. When used in the former manner they are called borers or drills; in the latter case they are of the form Fig. 25, and are termed jumpers. Recently power jumpers worked by compressed air and drills actuated in the same manner have been very successfully employed. Holes may be made by these instruments in almost any direction; but when hand labour only is available, the vertical can be most advantageously worked. Hand jumpers are usually about 4 feet 8 inches in length, and are used by holding in the direction of the required hole, and producing a series of sharp blows through lifting the tool about a foot high and dropping it with an impulsive movement. The bead divides a jumper into two unequal lengths, of which the shorter is used for commencing a bore-hole, and the longer for finishing it. Often the bit on the long length is made a trifle smaller than the other to remove any chance of its not following into the hole which has been commenced.

Drills and jumpers should be made of the best iron, preferably Swedish, for if the material be of an inferior quality it will split and turn over under the repeated blows of the mall, and thus endanger the hands of the workman who turns it, or give off splinters that may cause serious injury to those engaged in the shaft. Frequently they are made entirely of steel, and this material has much to recommend it for this purpose. The length of drills varies from 18 inches to 4 feet, the different lengths being put in successively as the sinking of the hole progresses. The cutting edge of the drills should be well steeled, and for the first, or 18-inch drill, have generally a

breadth of 2 inches; the second, or 28-inch drill, may be $1\frac{3}{4}$ inch on the edge; the third, or 3-foot drill, $1\frac{1}{2}$ inch, and the fourth, or 4-foot drill, $1\frac{1}{4}$ inch.

The mode of using the drill in the latter case is as follows: The place for the hole having been marked off with the pick, one man sits down holding the drill in both hands between his legs. Another man then strikes the drill with a mall, the former turning the drill partially round between each blow to prevent the cutting edge from falling twice in the same place.

The speed with which holes may be sunk varies of course with the hardness of the rock and the diameter of the hole. At Holyhead the average work done by three men in hard quartz rock with $1\frac{1}{2}$-inch drills was 14 inches an hour; one man holding the drill, and two striking. In granite of good quality, it has been ascertained by experience that three men are able to sink with a 3-inch jumper 4 feet in a day; with a $2\frac{1}{2}$-inch jumper, 5 feet; with a $2\frac{1}{4}$ inch, 6 feet; with a 2-inch, 8 feet; and with a $1\frac{3}{4}$-inch, 12 feet. A strong man with a 1-inch jumper will bore 8 feet in a day. The weight of the hammers used with drills is a matter deserving attention; for if too heavy they fatigue the men, and consequently fewer blows are given and the effect produced lessened; while, on the other hand, if too light, the strength of the workman is not fully employed. The usual weight is from 5 to 7 lb.

As the labour of boring a shot-hole in a given kind of rock is dependent on the diameter, it is obviously desirable to make the hole as small as possible, due regard being had to the size of the charge; for it must be borne in mind in determining the diameter of the boring that the charge should not occupy a great length in it. Various expedients have been resorted to for the purpose of enlarging the hole at the bottom so as to form a chamber for the powder. If this could be easily effected, such a mode of placing the charge would be highly advantageous, as a very small bore-hole would be sufficient, and the difficulties of tamping much lessened. One of these expedients is to place a small charge at the bottom of the bore, and to fire it after being properly tamped. The charge being insufficient to cause fracture, the parts in immediate contact with it are compressed and crushed to dust, and the cavity is thereby enlarged. The proper charge may then be inserted in the chamber thus formed by boring through the tamping. Another method, applicable chiefly to calcareous rock, was employed, it is stated, with satisfactory results, at Marseilles many years ago. When the bore-hole has been sunk to the required depth, a copper pipe, Fig. 26, of a diameter to fit the bore loosely, is introduced, the end A reaching to the bottom of the hole, which is closed up tightly at B with clay so that no air may escape. The pipe is provided with a bent neck, C. A small leaden pipe, e, about ½ inch in diameter, with a funnel, f, at the top, is introduced into the copper pipe at D, and passed to within an inch of the bottom. The annular space between the leaden and copper pipes is filled at g with hemp packing. Dilute nitric acid is then poured through the funnel and leaden pipe. The acid dissolves the calcareous rock at the bottom, causing effervescence,

Fig. 26.

and a slime containing the dissolved lime is forced out of the opening C. This process is continued until, from the quantity of acid used, it is judged that the chamber is enlarged sufficiently. Other acids would produce similar effects, the results in each case depending, of course, upon the chemical composition of the rock. The writer is unacquainted with any instance where this system has been used ; it is impracticable except in a few very special cases, and must even then be both troublesome and expensive.

After the shot-hole has been bored, it is cleaned out and dried with a wisp of hay, and the powder poured down ; or, when the hole is not vertical, pushed in with a wooden rammer. The quantity of powder should always be determined by weight. One pound, when loosely poured out, will occupy about 30 cubic inches, and 1 cubic foot weighs 57 pounds. A

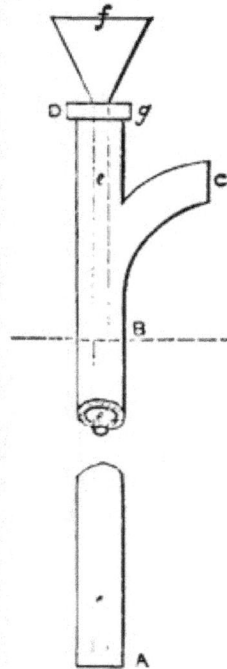

hole 1 inch in diameter will therefore contain ·414 ounce for every inch of depth. Hence to find the weight of powder to an

Fig. 27.   Fig. 29.

inch of depth in any given hole, we have only to multiply ·414 ounce by the square of the diameter of the hole in inches, and we are enabled to determine either the length of hole for a given charge, or the charge in a given space. It is important to use strong powder in blasting operations, because, as a smaller quantity will be sufficient, it will occupy less space and thereby save labour in boring.

When the hole is in wet stone, means must be provided for keeping the powder dry. For this purpose, tin cartridges are sometimes used. These are tin cylinders of suitable dimensions, fitted with a small tin stem through which the powder is ignited. The effect of the powder is, however, much lessened by the use of these tin cases. Generally a paper cartridge, well greased to prevent the water from penetrating, will give far more satisfactory results. When the paper shot is used, the hole should, previous to the insertion of the charge, be partially filled with stiff clay, and a round iron bar, called a clay-iron or bull, Figs. 27, 28, driven down to force the clay into the interstices of the rock through which the water enters. By this means the hole will be kept comparatively dry. The bull is withdrawn by placing a bar through the eye near the top of the former, provided for that purpose, and lifting it straight out. The cartridge is placed upon the point of a pricker and pushed down the hole. The pricker, shown

Fig. 28.

in Fig. 29, is a taper piece of metal, usually of copper to prevent accidents, pointed at one end and having a ring at the other. When the cartridge has been placed in its position by this means, a little oakum is laid over it, and a

Bickford fuse inserted. This fuse is inexpensive, very certain in its effects, not easily injured by tamping, and is unaffected by moisture. The No. 8 fuse is preferred for wet ground; and when it is required to fire the charge from the bottom in deep holes, No. 18 is the most suitable.

When the line of least resistance has been decided upon, care must be taken that it remains the line of least resistance; for if the space in bore-hole is not properly filled, the elastic gases may find an easier vent in that direction than in any other. The materials employed to fill this space are, when so applied, called tamping, and they consist of the chips and dust from the sinking, sand, well-dried clay, or broken brick or stones. Various opinions are held concerning the relative value of these materials as tamping. Sand offers very great resistance from the friction of the particles amongst themselves and against the sides of the bore-hole; it may be easily applied by pouring it in, and is always readily obtainable. Clay, if thoroughly baked, offers a somewhat greater resistance than sand, and, where readily procurable, may be advantageously employed.

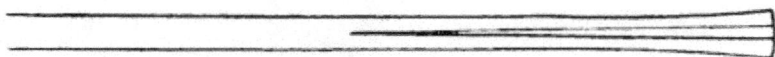

Fig. 30.   Fig. 31.

Broken stone is much inferior to either of these substances in resisting power. The favour in which it is held by sinkers and quarrymen, and the frequent use they make of it as tamping, must be attributed to the fact of its being always ready to hand, rather than to any excellent results obtained from its use. The tamping is forced down with a stemmer or tamping bar similar to Figs. 30, 31, too frequently made of iron, but which should be either of copper or bronze. The tamping end of the bar is grooved on one side, to admit of its clearing the pricker, or the fuse, lying along the side of the hole. The other end is left plain for the hand or for being struck with a hammer.

All tamping should be selected for its freedom from particles likely to strike fire, but it must not be overlooked that the cause of such a casualty may lie in the sides of the hole itself. Under these circumstances is seen the advisability of using bronze or copper tamping tools, and of not hammering violently on the tamping until a little of it has been first gently pressed down to cover over the charge, because the earlier blows on the tamping are the most dangerous in the event of a spark occurring. A little wadding, tow, paper, or a wooden plug is sometimes put to lie against the charge before any tamping is placed in the hole.

To lessen the danger of the tamping being blown out, plugs or cones of metal of different shapes are sometimes inserted in the hole. The best forms of plug are shown in Figs. 32 and 33; Fig. 32 is a metal cone wedged in on the tamping with arrows, and Fig. 33 is a barrel-shaped plug.

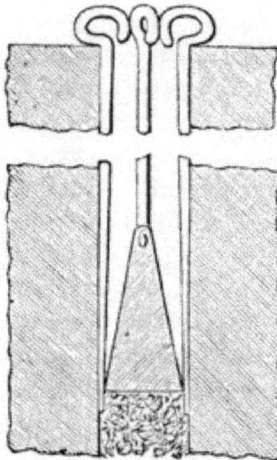

Fig. 32.

When all is ready, the sinkers, with the exception of one man whose duty it is to fire the charge, are either drawn out of the shaft, or are removed to some place of safety. This man then, having ascertained by calling and receiving a reply that all are under shelter, applies a light to the fuse, shouts "Bend away," or some equivalent expression, and is rapidly drawn up the shaft.

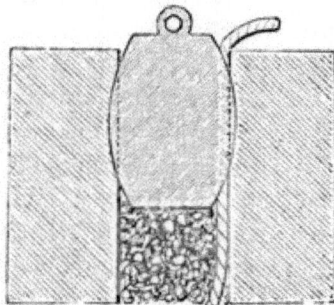

Fig. 33.

To avoid shattering the walls of a shaft, no shot should be placed nearer the side than 12 inches. The portion of stone next the wall sides of the shaft left after blasting is removed by steel-tipped iron wedges 7 or 8 inches in length. The wedges are applied by making a small hole with the point of the pick and driving them in with a mall. The sides may then be dressed as required with the pick.

After some 30 or 40 feet have been sunk the air at the

bottom of the well may be very foul, especially in a well
where blasting operations are being carried on, or where
there is any great escape of noxious gases through fissures.
Means must then be provided for applying at the surface a
small exhaust fan to which is attached lengths of tubing
extending down the well. Another good plan is to pass a 4 or
6 inch pipe down the well, bring it up with a long bend at
surface, and insert a steam jet; a brick chimney is frequently
built over the upper end of the pipe to increase the draught,
and the lower end continued down with flexible tubing. With
either fan or steam jet, the foul air being continuously withdrawn, fresh air will rush down in its place. This is far better
than dashing lime-water down the well, using a long wooden
pipe with a revolving caphead, or pouring down a vertical pipe
water which escaped at right angles, the old expedients for
freshening the air in a well.

A means of increasing the yield of wells, which is frequently
very successful, is to drive small tunnels or headings from the
bottom of the well into the surrounding water-bearing stratum.

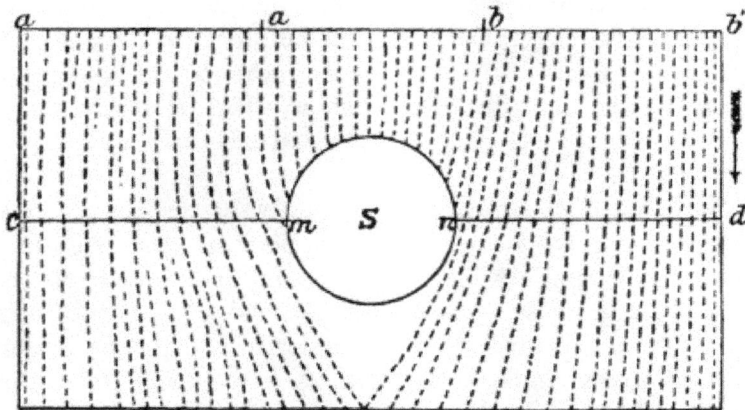

Fig. 34.

As an example, let Fig. 34 represent a sectional plan of a
portion of the water-bearing stratum at the bottom of the shaft.
This stratum is underlaid by an impervious stratum, and, consequently, the water will flow continuously through the former
in the direction of the dip, as shown by the arrow and the

dotted lines. That portion of the stratum to the rise of the shaft, S, which is included within vertical lines tangent to the circle at the points m and n, will be drained by the shaft. The breadth of this portion will, however, be extended beyond these lines by the relief to the lateral pressure afforded by the shaft, which relief will cause the fillets of water to diverge from their original course, towards the shaft, as shown in the figure. Hence the breadth of drainage ground will be $a\,b$, and it is evident that the shaft, S, can receive only that water which descends towards it through the space. But if tunnels be driven from the shaft along the strike of the stratum, as $m\,c$, $n\,d$, these tunnels will obviously intercept the water which flows past the shaft. By this means the drainage ground is extended from $a\,b$ to $a'\,b'$, and the yield of the well proportionately increased.

It should be remarked that when the stratum is horizontal or depressed in the form of a basin, that is, when it partakes more of the character of a *reservoir* than a *stream*, the only use of tunnels is to facilitate the ingress of water into the shaft, and in such case they should radiate from the shaft in all directions. They are also of service in case of accident to the pumps, as the time they take to fill up allows of examination and repairs being made in that time to the pumps, which could not be got at if the engines stopped pumping and the water rose rapidly up the shaft.

The size of the headings is usually limited by the least dimensions of the space in which miners can work efficiently, that is, about $4\frac{1}{2}$ feet high and 3 feet wide. The horse-shoe form is generally adopted for the sides and top, the floor being level, for the drawing off of the water by the pumps is quite sufficient to cause a flow, unless of course the dip of the stratum in which the tunnels are driven is such as to warrant an inclination. Where there is any water it is not possible to drive them with a fall, for the men would be drowned out.

The cost of some headings in the new red sandstone which the writer recently inspected, varied from 30s. a yard in ordinary stone, to 4l. 10s. a yard in very hard stone.

The foregoing remarks do not apply to headings driven in the chalk, where it is the usual practice to select the largest feeder issuing from a fissure and follow that fissure up, unless the heading is merely to serve as a reservoir, when the direction is immaterial.

The sides of wells usually require lining or steining, as it is termed, with some material that will prevent the loose strata of the sides of the excavation falling into the well and choking it. The materials that have been successfully used in this work are brick, stone, timber, and iron. Each description of material is suitable under certain conditions, while in other positions it is objectionable. Brickwork, which is universally used in steining wells in England, not unfrequently fails in certain positions; through admitting impure water when such water is under great pressure, or from the work becoming disjointed from settlement due to the draining of a running sand-bed, or the collapse of the well. Stone of fair quality, capable of withstanding compressive strains, is good in its way; but inasmuch as it requires a great deal of labour to fit it for its place, it cannot successfully compete with brickwork in the formation of wells, more especially as it has no merits superior to those of brick when used in such work; however, if in any locality, by reason of its cheapness, it can be used, care should be taken to select only such as contains a large amount of silica; indeed, in all cases it is a point of great importance in studying the nature of the materials used in the construction of wells, to select those which are likely to be the most durable, and at the same time preserve the purity of the water contained in the well; and this is best secured by silicious materials.

Timber is objectionable as a material to be used in the lining of wells, on account of its liability to decay, when it not only endangers the construction of the well, but also to some extent fouls the water. It is very largely used under some circumstances, especially in the preliminary operations in sinking most wells. It is also successfully used in lining the shafts of the salt wells of Cheshire, and will continue entire in such a position for a great number of years, as the brine seems to have a

tendency to preserve the timber and prevent its decay. Iron is
of modern application, and is a material

Fig. 35.

Fig. 36.

Fig. 37.

extensively employed in steining wells;
and, as it possesses many advantages over
materials ordinarily used, its use is likely
to be much extended. It is capable of
bearing great compressive strains, and of
effectually excluding the influx of all such
waters as it may be desirable to keep out,
and is not liable to decay under ordinary
circumstances. Baldwin Latham mentions
instances in his practice where recourse
has been had to the use of iron cylinders,
when it was found that four or five rings
of brickwork, set in the best cement, failed
to keep out brackish waters; and, if the
original design had provided for the introduction of these cylinders, it would have
reduced the cost of the well very
materially.

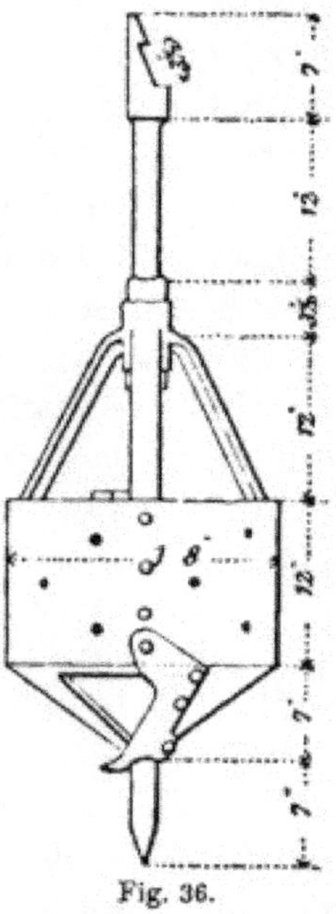

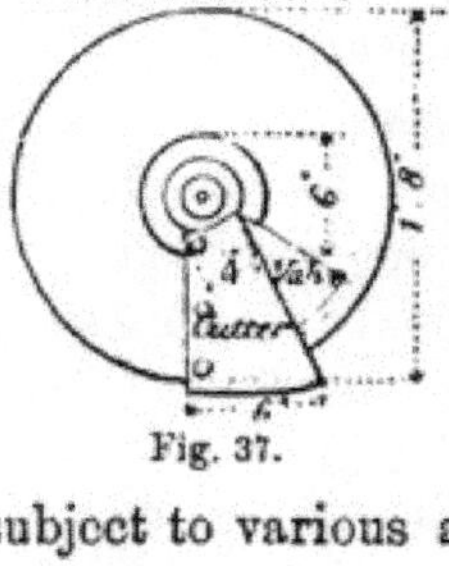

The well-sinker has often, in executing
his work, to contend with the presence of
large volumes of water, which, under ordinary circumstances, must be got rid of by
pumping; but by the introduction of iron
cylinders, which can be sunk under water,
the consequent expense of pumping is
saved.

When sinking these cylinders through
water-bearing strata, various tools are used
to remove the soil from beneath them. The
principal is the mizer, which consists of an
iron cylinder with an opening on the side
and a cutting lip, and which is attached to
a set of boring-rods and turned from above.

The valve in the old form of mizer is
subject to various accidents which interfere with the action of

the tool; for instance, pieces of hard soil or rock often lodge between the valve and its seat, allowing the contents to run out whilst it is being raised through water. To remedy this defect the eminent well-sinker, Thomas Docwra, designed and introduced the improved mizer, shown of the usual dimensions in Figs. 35 to 40; Fig. 35 being a plan at top, Fig. 36 an elevation, Fig. 37 a plan at bottom, Fig. 39 a section, Fig. 38 a plan of the stop *a*, and Fig. 40 a plan of the valve. It consists of an iron cylinder, conical shaped at bottom, furnished with holes for the escape of water, and attached to a central shank by means of stays. The shank extends some 7 inches beyond the bottom, and ends in a point, while the upper part of the shank has an open slot to form a box joint, Figs. 41 to 43, with the rods. The conical bottom of the mizer has a triangular-shaped

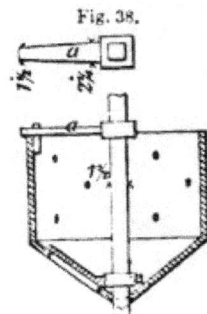

Fig. 38.

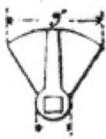

Fig. 39.

Fig. 40.

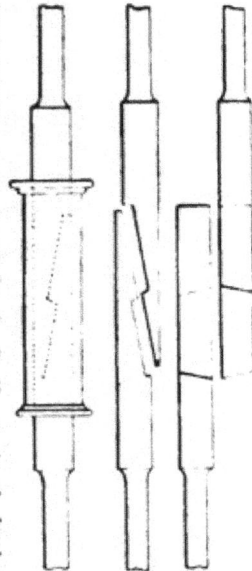

Fig. 41. Fig. 42. Fig. 43.

opening; on the outside of this is fitted a strong iron cutter, and on the inside a properly-shaped valve, seen in section and plan in Figs. 39 and 40. When the mizer is attached to and turned by means of the boring-rods, the *débris*, sand, or other soil to be removed, being turned up by the lip of the cutter, enters the cylinder, the valve, whilst the mizer is filling, resting against a stop. After the mizer is charged, which can be ascertained by placing a mark upon the last rod at surface and noting its progress downwards, the rods are reversed and turned once or twice in a backward direction; this forces the valve over the opening and retains the soil safely in the tool.

F

Fig. 44 is a pot mizer occasionally used in such soils as clay mixed with pebbles; there is no valve, as the soil is forced upwards by the worm on the outside, and falls over the edge into the cone.

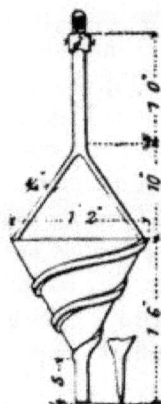

Mizers are fastened to the rods by means of the box joint, shown in Figs. 41 to 43, as a screw joint would come apart on reversing.

As many as five or six different sized mizers, ranging from 1 foot 6 inches to 9 feet in diameter, can be used successively, the smallest commencing the excavation, and the larger ones enlarging it until it is of the requisite size.

As an accessory, a picker, shown by the three views, Figs. 45 to 47, Fig. 46 indicating its correct position when in operation, is employed where the strata is too irregular or compact to be effectually cleared away by the cutter of the

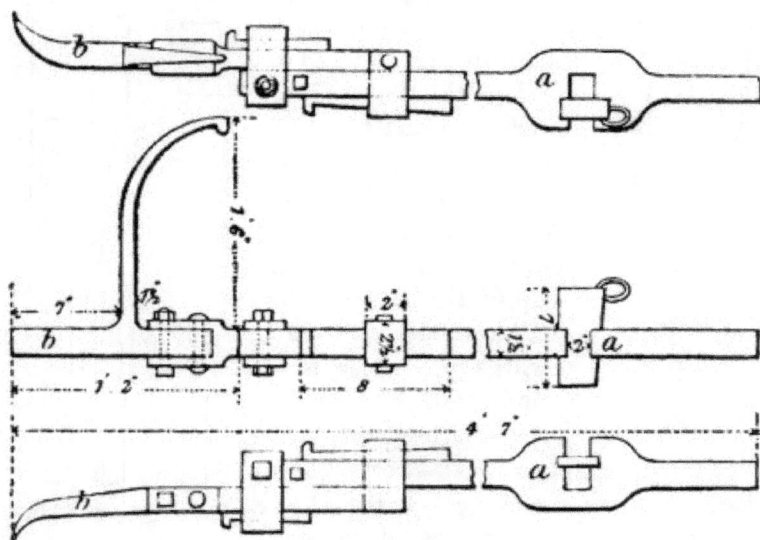

Fig. 44.

Figs. 45, 46, 47.

mizer. The picker is fixed upon the same rods above the

mizer, and is used simultaneously, being raised and lowered with that tool.

The cutting end of the picker is frequently replaced by a scratcher, Figs. 48, 49. This useful tool rakes or scratches up the *débris* thrown by the mizer beyond its own working range, and causes it to accumulate in the centre of the sinking, where it is again subjected to the action of the mizer.

Brick steining is executed either in bricks laid dry or in cement, in ordinary clay 9-inch work being used for large wells, and half-brick, or 4½-inch work, for small wells.

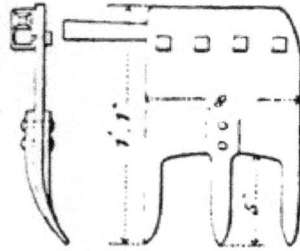

Figs. 48, 49.

Sectional Plan

Fig. 50.

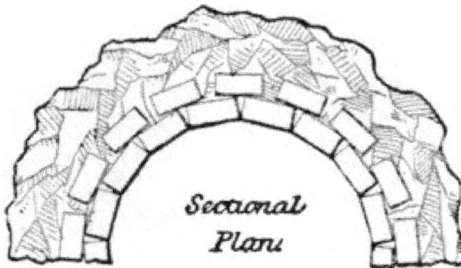

Sectional Plan

Fig. 51.

Figs. 50 and 51 show the method of laying for 9-inch work, and Fig. 52 for 4½ inches. The bricks are laid flat, breaking joint; and to keep out moderate land-springs clay, puddle, or concrete is introduced at the back of the steining; for most purposes concrete is the best, as, in addition to its impervious character, it adds greatly to the strength of the steining.   A

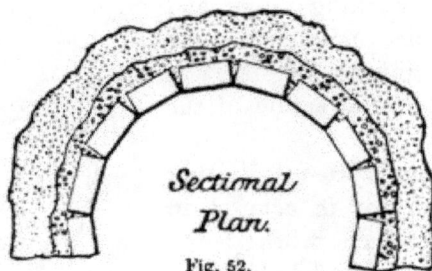

Sectional Plan.

Fig. 52.

ring or two of brickwork in cement is often introduced at intervals, varying from 5 feet to 12 feet apart, to strengthen the shaft, and facilitate the construction of the well.

Too much care cannot be bestowed upon the steining; if properly executed it will effectually exclude all objectionable infiltration, but badly made, it may prove a permanent source of trouble and annoyance. Half the wells condemned on account of sewage contamination really fail because of bad steining.

CHAPTER V.

WELL BORING.

THE first method of well boring known in Europe is that called the Chinese, in which a chisel suspended by a rope and surrounded by a tube of a few feet in length, is worked up and down by means of a spring-pole or lever at the surface. The twisting and untwisting of the rope prevents the chisel from always striking in the same place; and by its continued blows the rock is pounded and broken. The chisel is withdrawn occasionally, and a bucket or shell-pump is lowered, having a hinged valve at the bottom opening upwards, so that a quantity of the *débris* becomes enclosed in the bucket, and is then drawn up by it to the surface; the lowering of the bucket is repeated until the hole is cleaned, and the chisel is then put to work again.

Fig. 53 is of an apparatus, on the Chinese system, which may be used either for hemp-rope or wire-rope, and which was originally made for hoop-iron. At A, Fig. 53, is represented a log of oak wood, which is set perpendicularly so deep in the ground as to penetrate the loose gravel and pass a little into the rock, and stand firm in its place; it is well rammed with gravel and the ground levelled so that the butt of the log is flush with the surface of the ground, or a few feet below. Through this log, which may be, according to the depth of loose ground, from 5 feet to 30 feet long, a vertical hole is bored by an auger of a diameter equal to that of the intended boring in the rock. On the top of the ground, on one side of the hole, is a windlass whose drum is 5 feet in diameter, and the cog-wheel which drives it 6 feet; the pinion on the crank axle is 6 inches. This windlass serves for hoisting the spindle or drill and is of a large diameter, in order to prevent short bends in the iron, which would soon make it brittle.

In all cases where iron, either hoop-iron or wire-rope, is used, the diameter of the drum of the windlass used must be sufficiently large to prevent permanent bend in the iron. On the opposite side of the windlass is a lever of unequal leverage, about one-third at the side of the hole, and two-thirds at the opposite side, where it ends in a cross or broad end when men do the work. The workmen, with one foot on a bench or platform, rest their hands on a railing, and work with the other foot the long end of the lever. In this way the whole weight of the men is made use of. The lift of the bore-bit is from 10 to 12 inches, which causes the men to work the treadle from 20 to 24 inches high. Below the

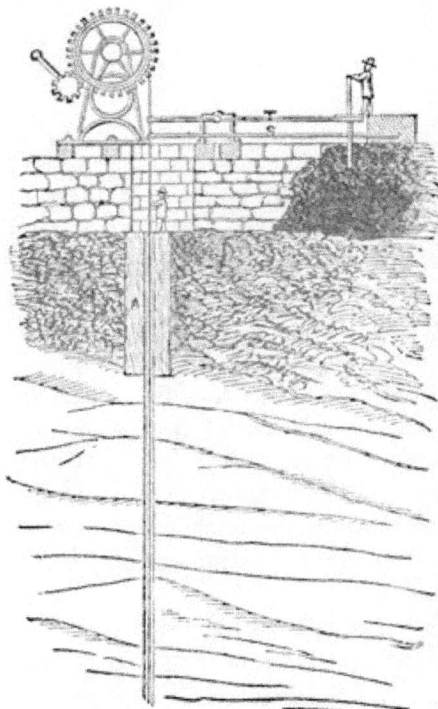

Fig. 53.

treadle, T, is a spring-pole, S, fastened under the platform on which the men stand ; the end of this spring-pole is connected by a link to the working end of the lever, or to the rope directly, and pulls the treadle down. When the bore-spindle is raised by means of the treadle, the spring-pole imparts to it a sudden return, and increases by these means the velocity of the bit, and consequently that of the stroke downwards.

This method has been generally disused, iron or wood rods substituted in the place of the rope, and a variety of augers and chisels instead of the simple chisel, with appliances for clearing the bore-hole of *débris*. Figs. 54 to 60 show examples of an ordinary set of well-boring tools. Fig. 56 is a flat chisel; Fig. 57 a V chisel; and Fig. 58 a T chisel. The flat chisel is for cutting up and loosening gravel and minerals that cannot be cut by the auger, the V and T chisels are for cutting into sandstone, limestone or other rock. These chisels are made from wrought iron, and when small are usually 18 inches long, $2\frac{1}{2}$ inches extreme breadth, and weigh some $4\frac{1}{2}$ lb.; the cutting edge being faced with the best steel. They are used for hard rocks, and whilst in operation need carefully watching that they may be removed and fresh tools substituted when their sides are sufficiently worn to diminish their breadth. If this circumstance is not attended to, the size of the hole decreases, so that when a new chisel of the proper size is introduced it will not pass down to the bottom of the hole, and much unnecessary delay is occasioned in enlarging it. In working with the chisel, the borer keeps the tiller, or handles, in both hands, one hand being placed upon each handle, and moves slowly round the bore, in order to prevent the chisel from falling twice, successively, in the same place, and thus preserves the bore circular. Every time a fresh chisel is lowered to the bottom it should be worked round in the hole, to test whether it is its proper size and shape; if this is not the case the chisel must be raised at once and worked gradually and carefully until the hole is as it should be. The description of strata being cut by the chisel can be ascertained with considerable accuracy by a skilful workman from the character of the shock transmitted to the rods.

When working in sandstone there is no adhesion of the rock to the chisel when drawn to the surface, but with clays the contrary is the case. Should the stratum be very hard, the chisel may be worn and blunt before cutting three-quarters of an inch, it must therefore be raised to the surface and frequently examined; however, 7 or 8 inches may be bored without examination, should the nature of the stratum allow of such progress being made.

Ground augers, Figs. 54, 55, and 60, are similar in action to those used for boring wood, but differ in shape and construc-

tion. The common earth or clay auger, Fig. 54, is 3 feet in
length, having the lower two-thirds cylindrical. The bottom is

Figs. 54,   55,   56,   57,   58,   59. .

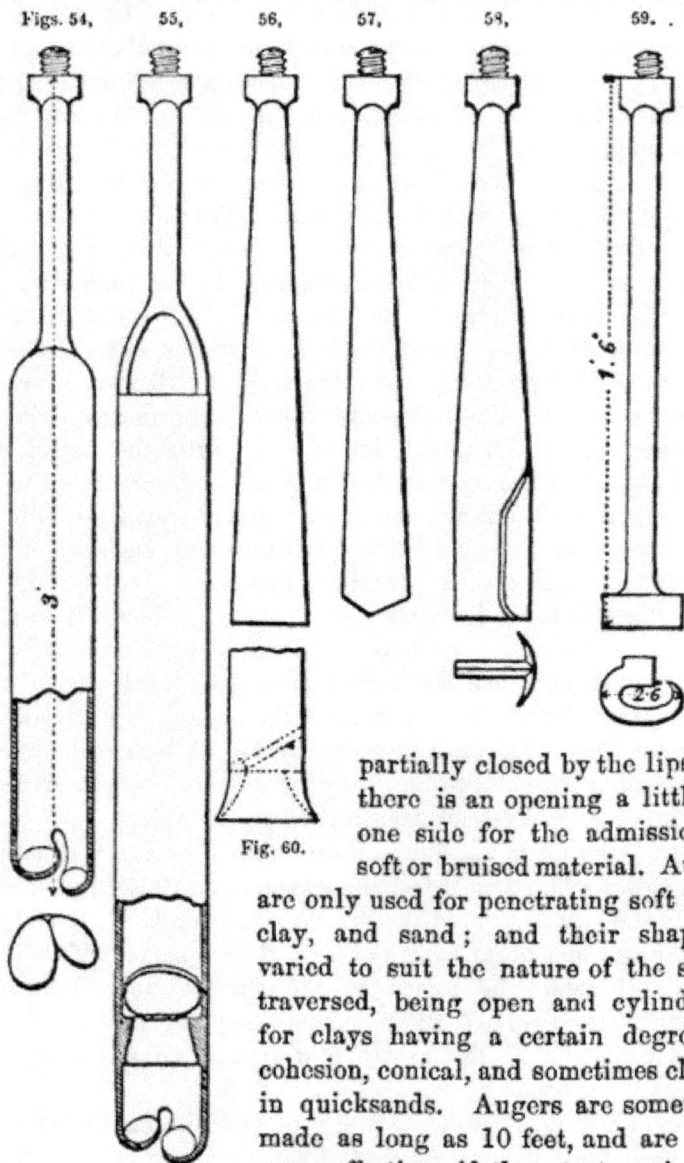

Fig. 60.

partially closed by the lips, and
there is an opening a little up
one side for the admission of
soft or bruised material. Augers
are only used for penetrating soft rock,
clay, and sand ; and their shape is
varied to suit the nature of the strata
traversed, being open and cylindrical
for clays having a certain degree of
cohesion, conical, and sometimes closed,
in quicksands. Augers are sometimes
made as long as 10 feet, and are then
very effective if the stratum is soft

enough to permit of their use. The shell is made from 3 feet
to 3½ feet in length, of nearly the same shape as the common
auger, sometimes closed to the bottom, Fig. 60, or with an
auger nose, Fig. 55; in either case there is a clack or valve
placed inside at the foot for the purpose of retaining borings of
a soft nature or preventing them from being washed out in a
wet hole. Fig. 63 shows a wad-hook for withdrawing stones, and
Fig. 62 a worm-auger.

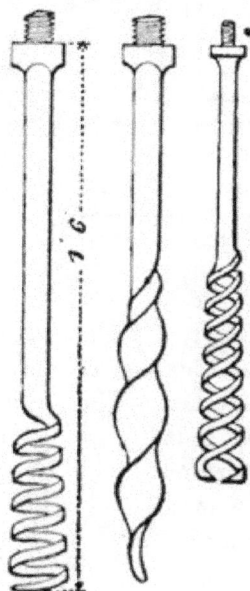

The Crow's Foot, Fig. 59, is used when the boring-rods have
broken in the bore-hole, for the purpose of extracting that por-tion
remaining in the hole; it is the same length, and at the foot
the same breadth as the chisels. When the rods have broken, the
part above the fracture is drawn out of the bore-hole and the
crow's foot screwed on in place of the broken piece; when this is lowered
down upon the broken rod, by careful twisting the toe is caused to grip the
broken piece with sufficient force to allow the portion below the fracture to

Fig. 64.

Figs. 61, 62, 63.

be drawn out of the bore-hole. A rough expedient is to fasten
a metal ring to a rope and lower it over the broken rod, when
the rod cants the ring, and thus gives it a considerable grip;
this is often very successful. Fig. 61 is a worm used for the
same purpose. A bell-box, Fig. 64, is frequently employed for
drawing broken rods; it has two palls fixed at the top of the
box, which rise and permit the end of the rod to pass when the
box is lowered, but upon raising it the palls fall and grip the
rod firmly. The action of these palls is rendered more certain
if they are arranged with flat springs pressing upon their upper
surface. A spiral angular worm, similar to Fig. 61, is also
applied for withdrawing tubes.

Of these withdrawing tools the crow is the safest and best, as it may be used without that intelligent supervision and care absolutely necessary with the worms and wad-hooks, or the bell-box.

The boring-rods (Figs. 65, 66) are in 3, 6, 10, 15, or 20 feet lengths, of wrought iron, preferably Swedish, and are made of different degrees of strength according to the depth of the hole

Fig. 65.

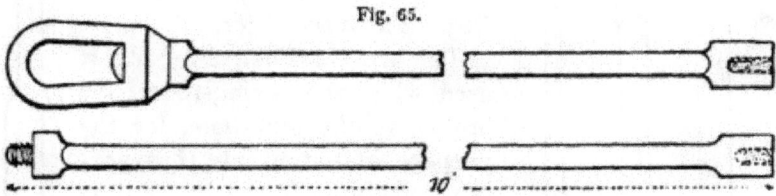

Fig. 66.

for which they are required; they are generally 1 inch square in section: at one end is a male and at the other end a female screw for the purpose of connecting them together. The screw should not have fewer than six threads. One of the sides of the female screw frequently splits and allows the male screw to be drawn out, thus leaving the rods in the hole. By constant wear, also, the screw may have its thread so worn as to become liable to slip. Common rods, being most liable to accident, should be carefully examined every time they are drawn out of the bore-hole, as an unobserved failure may occasion much inconvenience, and even the loss of the bore-hole. In addition to the ordinary rods there are short pieces, varying from 6 inches to 2 feet in length, which are fixed to the top, as required, for adjusting the rods at a convenient height.

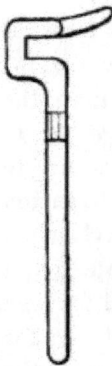

Fig. 67.

Fig. 68.

Fig. 69.

Fig. 67 is a hand-dog; Figs. 68 and 69, a lifting dog; Fig. 70, the tillers or handles by which the workmen impart a rotary motion to the tools. The tillers are clamped to the topmost

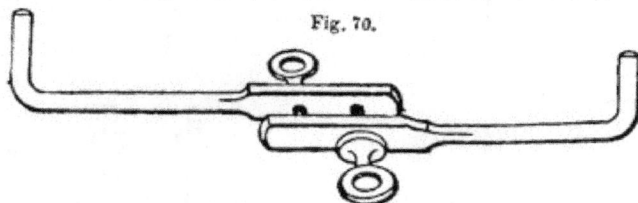

Fig. 70.

boring-rod at a convenient height for working. Fig. 65, a top rod with shackle. While the rods are sustained by the rope it allows them to turn freely. Fig. 71, a spring-hook for raising and lowering tools, rods, or pipes. It is spliced to a rope. When in use this should be frequently examined and kept in repair.

Lining tubes are employed to prevent the bore-hole falling in through the lateral swelling of clay strata, or when passing through running sand. The tubes are usually of iron, of good quality, soft, easily bent, and capable of sustaining an indent without fracture. Inferior tubes occasion grave and costly accidents, which are frequently irre-

Fig. 71.

parable, as a single bad tube may endanger the success of an entire boring.

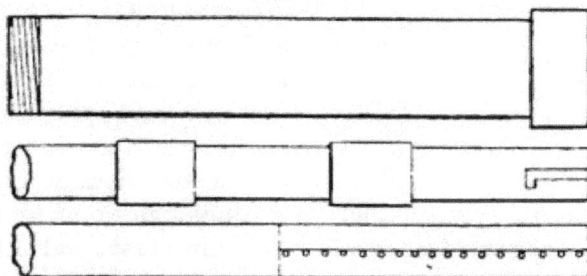

Fig. 72.

Figs. 73 and 74.

Wrought-iron tubes with screwed flush joints, Fig. 72, are to be recommended, but they are supplied brazed, Fig. 73, or

riveted, Fig. 74, and can be fitted with steel driving collars and
shoes.   Cast-iron tubes, Fig. 217, p. 161, are constantly applied ;
they should have turned ends with wrought-iron collars and
countersunk screws.   Care must be taken with socketed lining

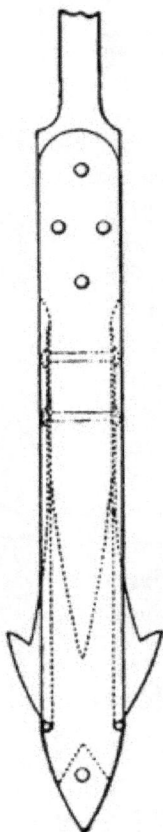

Fig. 75.

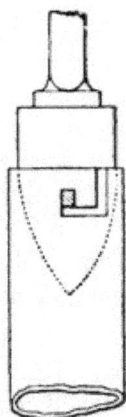

Fig. 76.

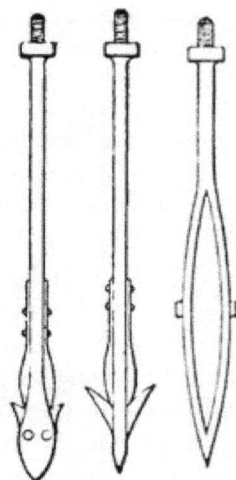

Fig. 77.   Fig. 78.   Fig. 79.

tubes to screw them together until they
butt in the centre of the socket, so as to
remove all strain on the screw threads
during the process of driving.

Cold-drawn wrought-iron tubes have
been used, and are very effective as well
as easily applied, but their relatively
high cost occasions their application to be
limited.

Fig. 76 shows a stud-block, which is used
for suspending tubing either for putting it down or for drawing
it up.   It consists of a block made to fit inside the end of the tube,
and attached to the rods in the usual way.   In the side of the
block is fixed an iron stud for slipping into a slot, similar to a
bayonet joint, cut in the end of the tube, so that it may be thus
suspended.   Figs. 75 and 77, 78, show various forms of spring-

darts, and Fig. 79 a pipe-dog, for the same purpose. Sometimes
a conical plug, with a screw cut around the outside for tightening itself in the upper end of the tube, is used for raising

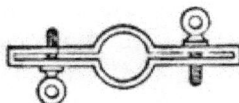

Fig. 80.

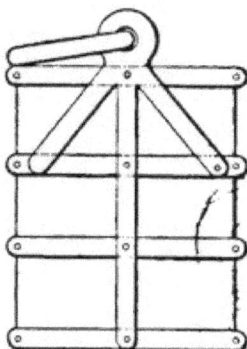

Fig. 81.

Fig. 82.

and lowering tubing. Figs. 80 and 81 are of tube clamps,
and Fig. 83 tongs for screwing up the tubes. Fig. 82 is of
an ordinary form of sinker's bucket.

Fig. 83.

Fig. 84 is a pipe-dolly, used for driving the lining tubes; the
figure shows it in position ready for driving.

When a projection in the bore-hole obstructs the downward
course of the lining tubes, the hole can be enlarged below the
pipes by means of a rimer, Fig. 85. It consists of an iron
shank, to which two thin strips are bolted, bowed out into the
form of a drawing pen. The rimer is screwed on to the boring-
rods, and forced down through the pipes; when below the last
length of pipe the rimer expands, and can then be turned
round, which has the effect of scraping the sides and enlarging
that portion of the hole subject to its operation. Fig. 86 is of
an improved form of rimer, termed a riming spring. It will be
seen that this instrument is much stronger than the ordinary

rimer, in consequence of the shank being extended through its entire length, thus rendering the scraping action of the bows very effective, whilst the slot at the foot of the bow permits of its introduction into, and withdrawal from, the tubing. Some means of suspending the tackle from which the rods are hung and also of obtaining a lift for them must be provided. Triangle gyns are sufficient for light work, whilst for that of a

Fig. 84.

Fig. 85.

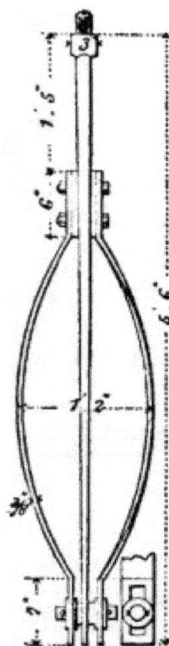

Fig. 86.

heavier character shears, derricks, or massive sheer-frames are requisite. Fig. 87 is a very good form of iron gyn for boring; it is of wrought-iron, and is fitted with a geared windlass.

In England, for small works, the entire boring apparatus is frequently arranged as in Fig. 88, the tool being fixed at the end of the wrought-iron rods instead of at the end of a rope, as in the Chinese method. Referring to Fig. 88, A is the boring tool; B the rod to which the tool is attached; D D the levers by which the men E E give a circular or rotating motion to the tool; F, chain for attaching the boring apparatus to the pole G,

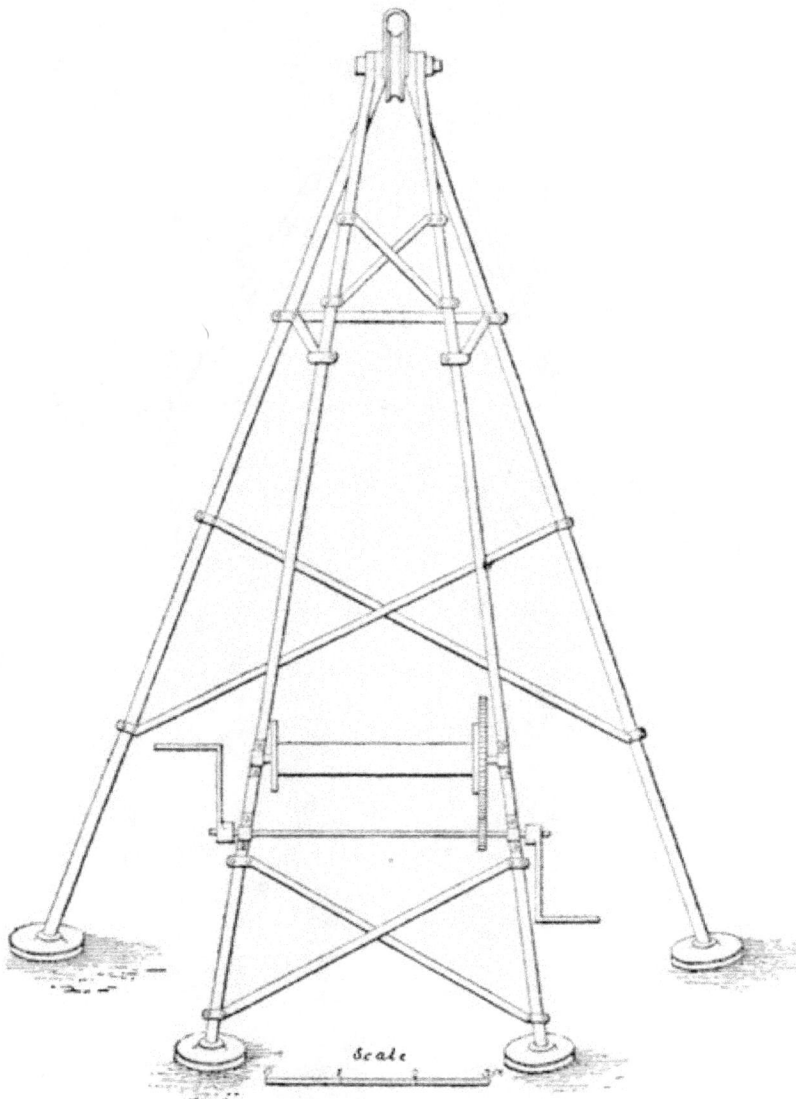

scale

Fig 87.

which is fixed at H, and by its means the man at I transmits a
vertical motion to the boring tool when this is necessary.

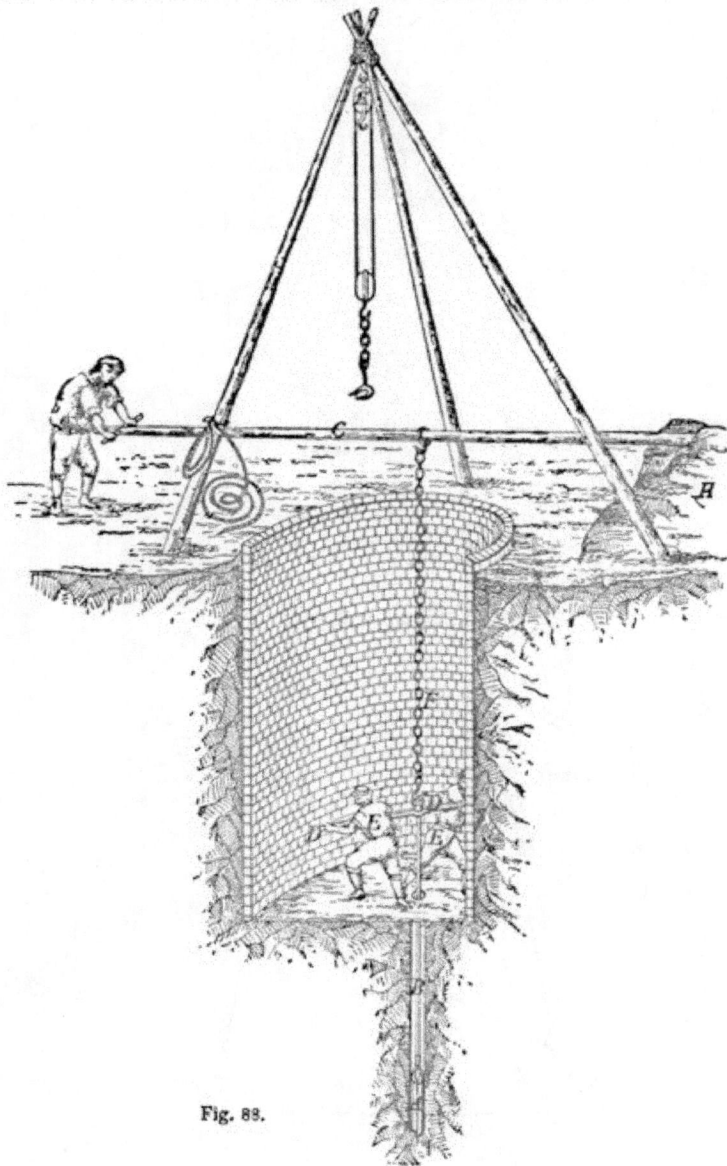

Fig. 83.

The sheer-legs, made of sound Norway spars not less than 8 inches diameter at the bottom, are placed over the bore-hole for the purpose of supporting the tackle K K for drawing the rods out of or lowering them into the hole, when it is advisable to clean out the hole or renew the chisel. It is obvious that the more frequently it is necessary to break the joints in drawing and lowering the rods, the more time will be occupied in changing the chisels, or in each cleaning of the hole, and as the depth of the hole increases the more tedious will the operation be. It therefore becomes of much importance that the rods should be drawn and lowered as quickly as possible, and to attain this end as long lengths as practicable should be drawn at each lift. The length of the lift or off-take, as it is termed, depending altogether upon the height of the lifting tackle above the top of the bore-hole, the length of the sheer-legs for a hole of any considerable depth should not be less than 30 to 40 feet ; and they usually stand over a small pit or surface-well, which may be sunk, where the clay or gravel is dry, to a depth of 20 or 30 feet. From the bottom of this pit the bore-hole may be commenced, and here will be stationed the man who has charge of the bore-hole while working the rods.

Fig. 89 is of another plan of commencing a boring. Here a, a are foot-blocks for the legs of the gyn, b the rope shackle, c d staging, e guide block. A pit is sunk 10 or 12 feet in the clear, when lined with timber or masonry, and below this a smaller pit 6 feet square, and 5 feet deep, also lined. Above these the sheer-legs are erected so that the rope when passed round the wheel at top may hang over the centre of the pits. The top of the lower part has to be covered, all except a gap of 9 inches in the centre, with loose planks to form a stage ; the two middle planks should be from 3 inches to 4 inches thick, as they may have to carry an auger board, and sustain the whole weight of the rods.

The arrangement, Fig. 90, is intended for either deep or difficult boring with rods. A regular scaffolding is erected, upon which a platform is built. The boring chisel A is, as in the last

G

instance, coupled by means of screw couplings to the boring rods
B. At each stroke two men stationed at E E turn the rod slightly
by means of the tiller D D. A rope F, which is attached to the

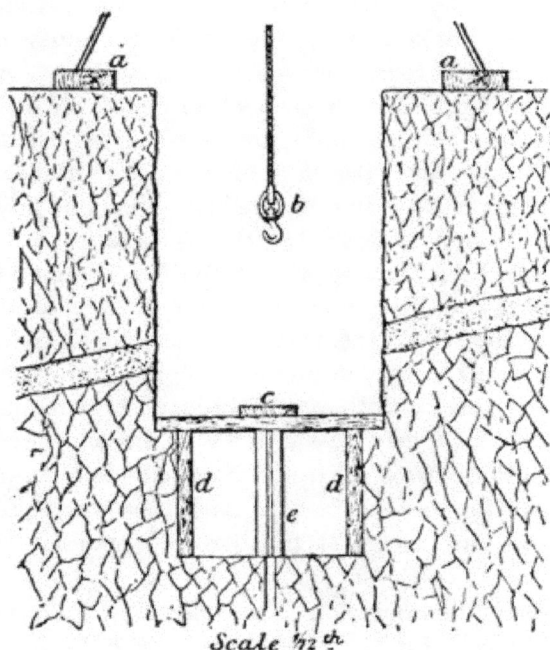

Scale ½₂ᵗʰ

Fig. 89.

boring tool, is passed a few times round the drum of a windlass
G, the end of the rope being held by a man at I. When the
handles are turned by the men at L L the man at I pulls at
the rope, the friction between the rope and the drum of the
windlass is then sufficient to raise the rods and boring tool, but
as soon as the tool has been raised to its intended height the
man at I slackens his hold upon the rope, and as there is insufficient friction on the drum to sustain the weight of the boring
tools, they fall. By a repetition of this operation the well is
bored, and after it has been continued a sufficient length of
time the tiller is unscrewed, and a lifting-dog, attached to the

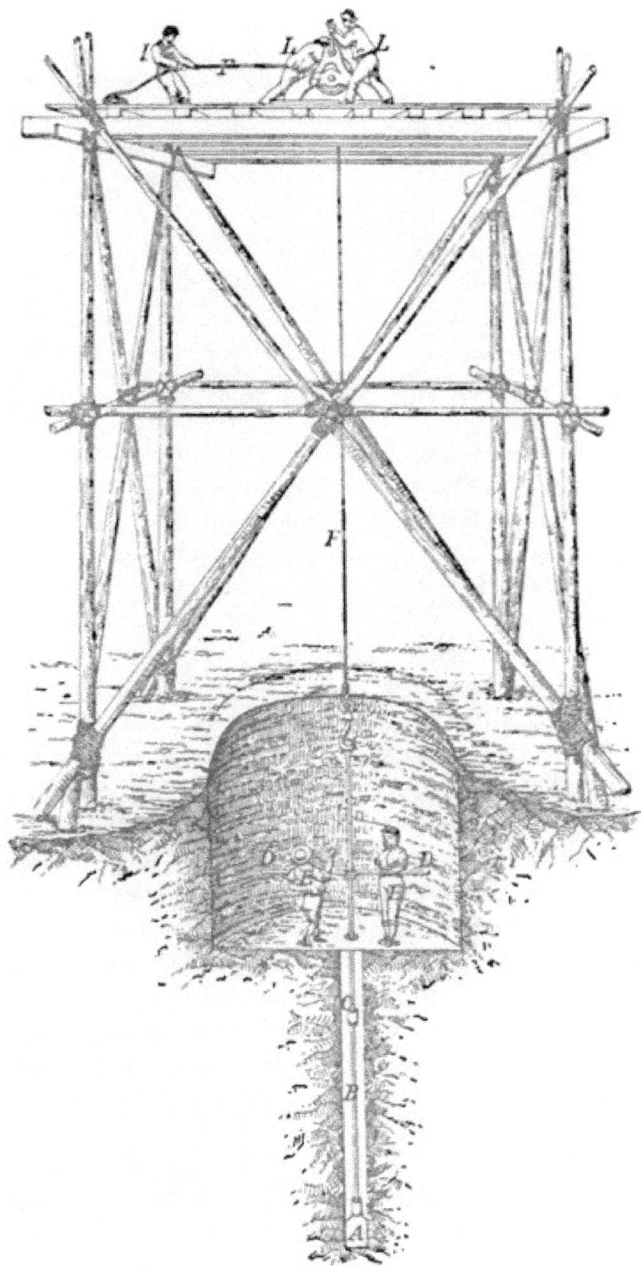

Fig. 90.

rope from the windlass, draws up the rods as far as the height
of the scaffolding or sheer-legs will allow, when a man at E,
Fig. 90, by passing a hand-dog or a key upon the top of the rod
under the lowest joint drawn above the top of the hole, takes
the weight of the rods at this joint, the men at L having lowered
the rods for this purpose; with another key the rods are unscrewed at this joint, the rope is lowered again, the lifting-dog
put over the rod, another rod screwed on, the rods lifted,
and the process continued until the chisel is drawn from the
hole and replaced by another, or, if necessary, replaced by some
other tool.

Sometimes if the hole is very dry, a little water poured
down assists the work, but care is necessary if the hole is still
unpiped, not to wash away the sides.

When a deep boring is undertaken, direct from the surface,
the operation had best be conducted with the aid of a boring
sheer-frame such as is shown in the frontispiece. This consists
of a framework of timber balks, upon which are erected four
standards, 27 feet in height, and 9 inches × 1 foot thick, 3 feet
8 inches apart at bottom, and 1 foot 2 inches at top, as seen in
the front and rear elevations. The standards are tied by means
of cross-pieces, upon which shoulders are cut which fit into
mortise holes, and are fastened by means of wooden keys, the
standards being surmounted by two head-pieces 5 feet long,
mortised and fitted. Upon the head-pieces two independent
cast-iron guide pulleys are arranged in bearings; over these
pulleys are led the ends of two ropes coiling in opposite directions upon the barrel of a windlass moved by spur gearing, and
having a ratchet stop attached to a pair of diagonal timbers, connected with the left-hand legs or standards of the sheers, near
the ground. These ropes are used for raising or lowering the
lengths of the boring rod.

Eight feet below the bearings of the top pulleys, a pair of
horizontal traverses are fixed across the frame, supporting
smaller pulleys, mounted on a cast-iron frame, which is capable
of motion between horizontal wooden slides. Over these
pulleys is led a rope from a plain windlass fixed to the right-

hand legs of the frame, to be used for raising or lowering the shell to extract the *débris* or rubbish from the hole.

The lever, 15 feet long, and 9 inches × 6 inches in section, is supported by an independent timber frame. It has a cast-iron cap, fastened by means of two iron straps, cast with lugs through which bolts are passed, these being tightened with nuts in the ordinary manner. The bearing pins at *a* are $1\frac{1}{2}$ inch in diameter, and also form part of the lower strap. Upon the cap is an iron hook, to this a chain is attached carrying a spring-hook which bears the top shackle of the rods. The top of the borehole is surrounded by a wooden tube 1 foot in diameter and surrounded by a hinged valve, whose action is similar to that of a clack-valve; this has a hole in the centre for the rods to pass up and down freely. The valve permits of the introduction and withdrawal of the tools, and at the same time prevents anything from above falling into the borehole.

The lever is applied by pressure upon its outer end, and as the relation of the long to the short arm is as 4 to 1, a depression of 2 feet in the one case produces an elevation of 6 inches in the other, the minimum range of action, the maximum being 26 inches.

With the sheer-frame the boring tools are worked in the same manner as in the preceding arrangements, Figs. 88, 90; but its portability, compactness, and adaptation of means to the required end, render its use desirable wherever it is possible to obtain it.

When in the progress of the work it is found that the auger does not go down to the depth from which it was withdrawn, after trial, tubing will generally be necessary. The hole should be enlarged from the surface, or, if not very deep, commenced afresh from the surface with a large auger, and run down to nearly the same depth; the first length of tube is then driven into the hole, and when this is effected, another tube, having similar dimensions to the first, is screwed into its upper end, and the driving repeated, and so on until a sufficient number of pipes have been used to reach to the bottom of the hole. If the ordinary auger is now introduced through these tubes it will

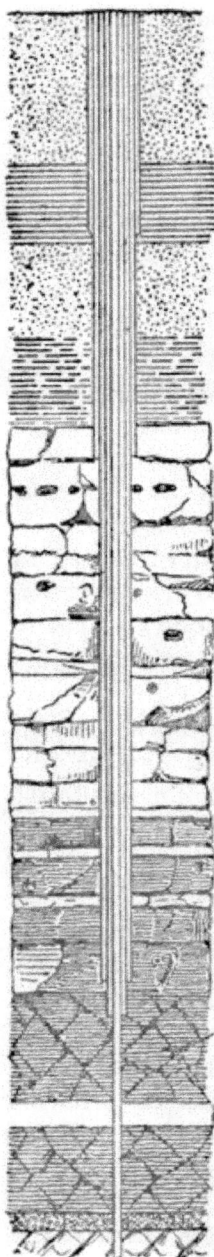

Fig. 91.

have free access to the clay or sand, and after a few feet deeper have been bored another pipe may be screwed on, and the whole driven farther down. In this way from 10 to 20 feet of soft stratum may be bored through. If the thickness of the surface clay or sand is considerable, the method here mentioned will not be effective, as the friction of the pipes, caused by the pressure of the strata, will be so great that perhaps not more than 80 or 100 feet can be driven without the pipes being injured. It will then be necessary to put down the first part of the borehole with a large auger, and drive in pipes of larger diameter; the hole is continued of smaller diameter, and lined with smaller tubes projecting beyond the large tubes, as in Fig. 91, until the necessity for their use ceases.

It will be evident that to ensure success the tubing, whatever it is made of, should be as truly cylindrical as possible, straight, and flush surface, both outside and in. It will also be evident that in thus joining pieces of tubing together, the thickness ought to have a due proportion to the work required, and the force likely to be used in screwing or driving them down. The first or bottom pipe is furnished with a steel shoe, having a chisel edge serving to trim the hole and cut a passage for the sockets of the tubing to pass freely. The first length of pipe is raised by means of a pipe hanger, and lowered into the borehole until its top reaches within 1 foot of the bottom of the pit; here a pair of pipe-clamps are securely fastened round it a few inches above the thread, and then the pipe is lowered until the clamps rest upon the board surrounding the top of the hole. The hanger is removed and screwed on to a fresh length of tubing, and this in its turn lowered and screwed quite home, until the two pipes butt together. The tillers being taken off, the whole length of tubing is raised a few inches and suspended whilst the clamps are removed from the lower part. There are now two lengths of pipe; they are allowed to descend as before; when they are sufficiently deep the clamps are reapplied, and the operation repeated with each length screwed on.

Each joint should be oiled and screwed together with white or red lead: spun yarn is not needed. Every socket must be removed and screwed on again with red lead before being attached to a fresh pipe.

While being lowered the pipes are turned, particularly when they begin to hang up, in order that the steel shoe may remove any projections in the borehole. When the pipes have been lowered to the necessary distance, and the pipe-clamps screwed on to secure them from slipping, boring can be resumed with the smaller sized boring tools, after lowering the shell to bring up any *débris* caused through lowering the tubing.

When it is necessary to get the tubing lower and it will not go down freely, the rimer, Fig. 85, may be employed if the stratum is not too hard for them to cut. It is screwed on to the bottom rod, and as the springs measure the outside diameter of the tubing, they require to be pressed so as to force them through, although when once well in the pipes the weight of the rods should be sufficient to carry them down. As soon as the springs are below the lowest length of pipe, they expand to their full size, and by turning the rods until the springs work quite freely, and lowering the rimer a little as they are freed, the hole below the tubing can then be cut out as wide as is necessary.

When the rimer has been withdrawn, the pipes are attached and lowered as before. It will be observed that using the rimer is an operation requiring great care and attention.

In the manner detailed above, the tubing should be turned as long as it will move before resorting to driving. It is desir-

able to use all the longer lengths of pipe first, reserving the
shorter lengths to the last, when the tubing will be going down
more slowly than at the first.   A long length standing up, at a
time when it is necessary to lower tools for clearing or
enlarging below the tubing, may seriously obstruct the work.
Sometimes a short length of pipe may be used temporarily with
advantage, a few feet of the descent proceeded with, and then a
longer length can be substituted as soon as the boring has
proceeded sufficiently for a further lowering of the pipes.
Wrought-iron tubes, when driven, must be worked carefully, by
means of a ring made of wrought-iron, from $1\frac{1}{2}$ to 2 inches
in height, and $\frac{3}{4}$ inch thick, and of the form shown in Fig. 92 ;
or driven with a pipe dolly such as that in Fig. 84.   The ring,
or the dolly, is screwed into the lowermost boring-rod and
worked at the same rate and in a similar manner to the chisel,
due regard being had to the depth at which the driving is being
done, as the weight of the boring-rods will materially affect the
strength of the blow delivered.   Cast-iron tubing may be driven
hard with a monkey, or forced down by screw-jacks or hydraulic
pressure by the methods illustrated in Figs. 218, 221.

To withdraw broken or defective tubing quickly, two hooks
attached to ropes are lowered down from opposite sides of the
borehole, caught on the rim of the lowermost tube, and power
applied to haul the tubing up bodily.

Figs. 93 to 97 show good methods of forming tube or pipe
joints both in cast and wrought iron, when not screwed.

P. S. Reid, an English mining engineer, gives the following
instance of replacing defective tubing in a boring which had
been pursued to the depth of $582\frac{1}{2}$ feet, but which, owing
to circumstances which were difficult to determine, had become
very expensive, and made slow progress.

The $582\frac{1}{2}$ feet had been bored entirely by manual labour ;
but Reid recommended the erection of a horse-gyn in which the
power was applied to a 40-inch drum placed upon a vertical
axle, the arms of which admitted of applying two horses, and
men at pleasure, the power gained being in the proportion
of one to ten at the starting-point for the horses.

Upon the upright drum a double-ended chain was attached, which worked from sheer-legs erected immediately over the hole, so as to attain an off-take for the rods of 60 feet, and so

Fig. 93.

Fig. 94.

Fig. 95.

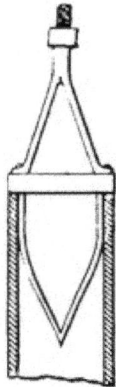

Fig. 92.

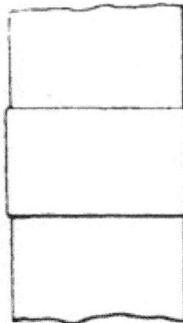

Fig. 96.

Fig. 97.

as that, in the act of raising or lowering, there might always be one end of the chain in the bottom, ready to be attached, and expedite the work as much as possible.

These arrangements being made, it was soon found that there was a defect in the tubing which was inserted to the depth of 109 feet, and the defect was so serious, in permitting the sand to descend and be again brought up with the boring tools, as to render it very difficult to tell in what strata they really were; this increased to such a degree as to cause the silting up of the hole in a single night to the extent of 180 feet, and it occupied nearly a fortnight in clearing the hole out again.

On carefully examining into this defect, it appeared that the water rose in the hole to the depth of 74 feet from the surface; and that at this point it was about level with the high-water mark on the river Tees, about two miles distant, with which it was no doubt connected by means of permeable beds, extending from the arenaceous strata at a depth of 100 feet.

On commencing to bore, the motion of the rods in the hole caused the vibration of the water between a range of 40 feet at the bottom of the tubing, and so disturbed the quiescent sand as to cause it to run down through the faults in the lower end of the tubing.

This tubing was made of galvanised iron plates, riveted together and soldered; at the top of the hole it was in three concentric circles which had been screwed and forced down successively until an obstacle was met with at three different places. So soon as the outer circle reached the first depth, all hope appears to have vanished, from those who bored the earlier part of the work, of getting the tube farther; a second tube was, therefore, inserted, which seems to have advanced as far as the second obstacle, where it, in its turn, was abandoned; and a third one advanced until it rested in the strata at the lower part of the lias freestone of a blue nature, as found on the rocks at Seaton Carew, and in the bed of the Leven, near Hutton Rudby. The diameter of the first tubing was $3\frac{7}{8}$ inches external and $3\frac{1}{2}$ inches internal; the second tube was $3\frac{1}{4}$ inches external, and 3 inches internal diameter; and the third tube was $2\frac{3}{4}$ inches external and $2\frac{1}{2}$ inches internal diameter.

Such being the account gathered from the workmen who superintended the earlier part of the boring, it became necessary to decide upon the best course to remedy the evil. At first sight it would have appeared easy enough to have caught the lower end of the tubes by means of a fish-head properly contrived, and thus to have lifted them out of the hole, 'and replaced them

with a perfect tube, such as a gas-tube, with faucet screw joints;
but, on attempting this, it soon became evident that however
good the tubing which might have been adopted, it would be a
work of the greatest difficulty to extract when once it was
regularly fixed and jammed into its place by the tenacious
clayey strata surrounding it; and the difficulty of extracting it,
in the present case, was even enhanced by the inferior quality
and make of the tubing; in short, that, unless by crumpling it
up in such a manner as to destroy the hole, it was impossible to
extract this tubing by main force.

There was, therefore, no other choice left but to attempt
cutting it out, inch by inch; though before doing so, force was
applied to the bottom of the tubing, to the extent of upwards of
30 tons, the only result being the loss of several pieces of steel
down the hole, which had to be brought up with a powerful
magnet.

After much mature consideration and contrivance, it was
determined to order such tubing as would at the same time
present as little obstruction as possible to the clay to be passed
through on the outside, as well as surround the largest of the
three tubes then in the hole, and present no obstacle to their
being withdrawn through its interior.

These tubes were made 12 feet in length, flush outside and
in, the lower portion being steeled for 6 inches from the
bottom end, so as to cut its way and follow down the space, and
cover that exposed by the old tubes when cut and drawn, as
shown in Fig. 98.

In order to commence operations, and avoid too much clay
going down to the bottom of the hole, a straw-plug was firmly
fixed in the lias portion of the hole. The lower portion of the
new tubes was then screwed around the old ones by means of
powerful clamps, attached to the exterior in such a manner as to
avoid injuring the surface; and when they could be screwed no
farther, the knife or cutter, Figs. 98 to 100, was introduced inside
the old tubing. Some force was needed to get this knife down
into the tubing, but the spring *a* giving so as to accommodate

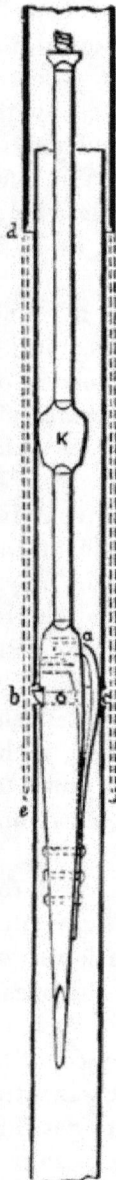

itself to the hole, permitted its descent to the distance required; this being effected, it was turned round so that the steel cutter, shown at *b*, being forced against the sides of the tube, cut it through in the course of ten minutes or a quarter of an hour's turning. See section at *b, c*, Fig. 99.

The old tubes being three-ply, three of these knives or cutters were required to cut out the three tubes, the inner one being detached first, and then the two exterior ones; and so soon as these latter were cut out as far as they had been forced into the clay, the work became simplified into following down the interior tubing by the new tubes, as shown by the dotted lines. From *d* at the lower end, it was found that the old inner tube had been so damaged or torn, either by the putting in or hammering it down, as to leave a vent or fissure for the sand to descend,

Fig. 100. and thus spoil the whole of the work for all future success in the boring, to say nothing of the very great cost of lifting the sand out, and subsequent most arduous labour to put the hole right.

Boring was recommenced after about a month's labour in taking out the old

Fig. 99.

Fig. 98.

STEELED 6 INS MONTHS END

tubings, leaving the new ones firmly bedded into the lias, 112 feet from the surface, and the hole was subsequently bored to a depth of 710 feet in the new red sandstone, proceeding at the rate of about 3 feet in the twelve hours, and leaving the hole so as, if requisite, it might be widened out to 4 inches diameter. Fig. 98 shows the action of the knife and spring-cutter when forced down into the tubing, ready to commence cutting. It also shows the lower end of the new tubing, enclosing the others at the commencement of the work. The joints of the new tubes were made by means of a half-lap screw. Fig. 100 is a back view of the knife or cutter b. Fig. 99 shows the action of the spring and cutter when the requisite length is cut through and ready for lifting; the position of the tube being maintained perpendicular, or nearly so, by the ball or thickening on the rods at K, and the lower end of the tube being supported by the projecting steel cutter at b, the dotted lines from d showing the position of the new steel-ended tube when screwed down ready for another operation. In boring deeper after the tubes were removed, three wooden blocks were used round the rods in the new tube to keep them plumb.

A more effective method of cutting out lining tubes is that practised in the United States. This consists in lowering into the borehole an expanding cutter-head, in which the circular cutters are first tightened, and then put into action by turning the boring-rods at surface.

To reduce the stoppages for the withdrawal of *débris* the system of Fauvelle was introduced, but it is now very little practised on the Continent, and not at all in Great Britain. The principles upon which it was founded were: first, that the motion given to the tool in rotation was simply derived from the resistance that a rope would oppose to an effort of torsion; and therefore that the limits of application of the system were only such as would provide that the tool should be safely acted upon; and, secondly, that the injection of a current of water, descending through a central tube, should wash out the *débris* created by the cutting tool at the bottom. The difficulties attending the removal of the *débris* were great; and though

the system of Fauvelle answered tolerably well when applied
to shallow borings, it was found to be attended with such
disadvantages when applied on a large scale, that it has been
generally abandoned. The quantity of water required to keep
the boring tool clear is a great objection to the introduction
of this system, especially as in the majority of cases Artesian
wells are sunk in such places as are deprived of the advantage
of a large supply.

In the ordinary system of well boring, innumerable breakages and delays occur when a boring is required to be carried to
any depth exceeding 200 or 300 feet, owing to the buckling of
the rods, the crystallisation of the iron by the constant jarring
at each blow, and particularly the increased weight of the rods
as the hole gets deeper. It follows from this, that where the
excavation is very deep, there is considerable difficulty in transmitting the blow of the tool, in consequence of the vibration
produced in the long rods, or in consequence of the torsion ;
and, for the same reason, there is a danger of the blows not
being equally delivered at the bottom. It has been attempted
to obviate this difficulty, but without much success, by the use
of hollow rods, presenting greater sectional area than was absolutely necessary for the particular case, in order to increase their
lateral resistance to the blows tending to produce vibration.

Boring is usually executed by contract. The approximate
average cost in England may be taken at 1s. 3d. a foot for the
first 30 feet ; 2s. 6d. a foot for the second 30 feet ; and continue
in arithmetical progression, advancing 1s. 3d. a foot for every
additional 30 feet in depth. This does not include the cost of
tubing, conveyance of plant and tools, professional superintendence, or working in rock of unusual hardness, such as hard
limestone and whinstone. A clause is usually inserted in the
contract, to the effect that, if any unforeseen difficulty is met
with in the course of the work, it is then paid for by the day,
at a rate previously determined upon, until the difficulty has
been overcome.

CHAPTER VI.

THE TUBE WELL.

THIS well consists of a hollow wrought-iron tube about $1\frac{3}{4}$ inch diameter, composed of any number of lengths, each from 3 to 11 feet, according to the depth required. The water is admitted into the tube through a series of holes, which extend up the lowest length to a height of $2\frac{1}{2}$ feet from the bottom. Common gas or other pipes are of no service for a tube well ; specially tough lap-welded tubes are necessary, in order to withstand the hammering and vibration to which they are subjected.

The position for a well having been selected, a vertical hole is made in the ground with a crow-bar to a convenient depth ; the well tube a, having the clamp d, monkey c, and pulleys b Fig. 101, previously fixed on it, is inserted into this hole.

The clamp is then screwed firmly on to the tube from 18 inches to 2 feet from the ground, as the soil is either difficult or easy to drive in ; the bolts being tightened equally, so as not to indent the tube.

The pulleys are next clamped on to the tube at a height of about 6 or 7 feet from the ground, the ropes from the monkey having been previously rove through them.

The monkey is raised by two men pulling the ropes at the same angle. They should stand exactly opposite each other, and work together steadily, so as to keep the tube perfectly vertical, and prevent it from swaying about while being driven. If the tube shows an inclination to

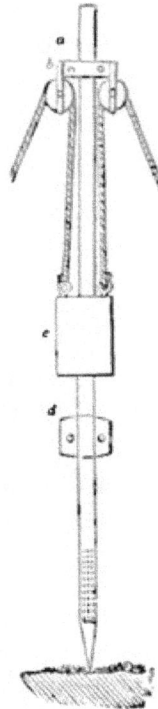

Fig. 101.

slope towards one side, a rope should be fastened to its top and
kept taut on the opposite side, so as gradually to bring the
tube back to the vertical. When the men have raised the
monkey to within a few inches of the pulleys, they lift their
hands suddenly, thus slackening the ropes and allowing the
monkey to descend with its full weight on to the clamp. The
monkey is steadied by a third man, who also aids to force it
down at each descent. This man, likewise, from time to time,
with a pair of pipe-tongs, turns the tube round in the ground,
which assists the process of driving, particularly when the
point comes in contact with stones.

Particular attention must be paid to the clamp, to see that it
does not move on the tube; the bolts must be tightened up at
the first appearance of any slipping.

When the clamp has been driven down to the ground, the
monkey is raised off it, the screws of the clamp are slackened,
and the clamp is again screwed to the tube, about 18 inches or
2 feet from the ground. After this, the monkey is lowered on
to it, and the pulleys are then raised until they are again 6 or
7 feet from the ground.

The driving is continued until but 5 or 6 inches of the well
tube remain above the ground, when the clamp, monkey, and
pulleys are removed, and an additional length of tube screwed
on to that in the ground. This is done by first screwing a
collar on to the tube in the ground, and then screwing the next
length of tube into the collar, till it butts against the lower
tube; a little white-lead must be placed on the threads of the
collar before the ends of the tubes are screwed into it.

The driving can thus be continued until the well has obtained
the desired depth. Soon after another length has been added,
the upper length should be turned round a little with the pipe-
tongs, to tighten the joints, which have a tendency to become
loose from the jarring of the monkey. Care must be taken,
after getting into a water-bearing stratum, not to drive through
it, owing to anxiety to get a large supply. From time to time,
and always before screwing on an additional length of tube, the
well should be sounded, by means of a small lead attached to a

line, to ascertaiu the depth of water, if any, and character of the
earth which has penetrated through the holes perforated in the
lower part of the well tube. As soon as it appears that the
well has been driven deep enough, the pump is screwed on to
the top and the water drawn up. It usually happens that the
water is at first thick, and comes in but small quantities; but
after pumping for some little time, as the chamber round the
bottom of the well becomes enlarged the quantity increases and
the water becomes clearer.

When sinking in gravel or clay the bottom of the well tube
is liable to become filled up by the material penetrating through
the holes; and before a supply of water can
be obtained, this accumulation must be
removed by means of the cleaning pipes.

The cleaning pipes are of small diameter,
½-inch externally, and the several lengths are
connected together in the same way as the
well tubes, by collars screwing on over the
adjoining ends of the two pipes.

To clear the well, one cleaning pipe after
another is lowered into the well, until the
lower end touches the accumulation; the pipes
must be held carefully, for if one were to
drop into the well it would be impossible to
get it out without drawing the well. A pump
is then attached to the upper cleaning pipe
by means of a reducing socket; the lower end
of the cleaning pipe is then raised and held
about an inch above the accumulation by
means of the pipe-tongs: water is next poured
down the well outside the cleaning pipe, and,
being pumped up through the cleaning pipe,
brings up with it the upper portion of the
accumulation; the cleaning pipe is gradually
lowered, and the pumping continued until the

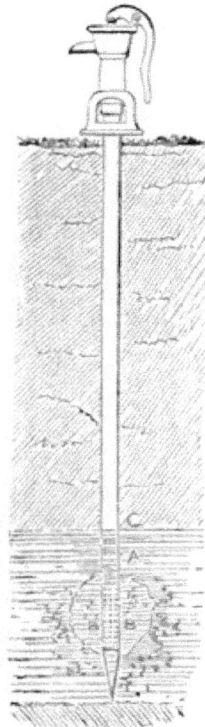

Fig. 102.

whole of the stuff inside the well tube is removed. The pump
is then removed from the cleaning pipe, and the cleaning pipes

H

are withdrawn piece by piece; and finally the pump is screwed on to the upper end of the tube well, Fig. 102, which is then in working order.

In practice it has been found that when driving in very hard strata, immediately the point has penetrated through them the driving becomes quite easy again, irrespective of the depth that may have been reached. For instance, it is common for a tube to take many hours driving through an obstruction of 3 feet or 4 feet, met with, say, at 20 feet or 30 feet below the surface, and subsequently driving at four times the pace when reaching, possibly, double or treble the depth. From this fact it is quite evident that the first tube, with its point, accomplishes the whole work of penetration, leaving the tubes that follow practically no resistance to overcome. Instead, therefore, of striking the tube at or about the surface, and having to transmit the blow through any length of pipe, varying according to the depth driven, in Le Grande and Sutcliff's arrangement the blow is delivered immediately above the point at the bottom of the tube.

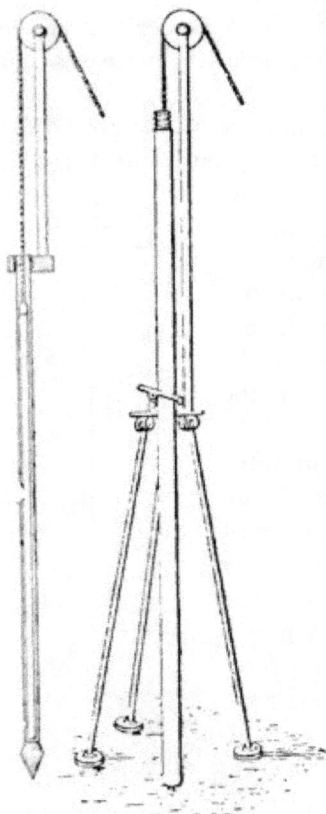

This is accomplished by using an elongated weight, shod with steel which passes down the inside of the tube. To this weight is attached a long rope, which is pulled by hand and allowed to fall, striking upon an even surface provided at the top of the bulbous point; this is the most simple form of working on this plan, and is shown in Fig. 103. Fig. 104 shows a more convenient form of work-

Fig. 103.   Fig. 104.

ing. A short sheer-legs is arranged so that the head serves as a guide for the tube, and from this head rise two tubular uprights, to carry a pulley, over which the rope travels, and which can be worked either by hand or power. One feature of this system is that it obviates the risk of bending the tubes when they get on to anything hard. This plan of internal driving is found to be very applicable for tubes of large diameter, as the tackle required is simple and inexpensive; it also is useful for driving under a head of water.

The smallest sized tube wells are advantageous for manufacturing purposes, for supplying boilers with feed water, not only in rural districts, but in towns supplied with waterworks, thus saving water rates: many are thus in use for similar purposes even in the very heart of London, springs being often found to exist in spots when it might have been expected that deep sewers had drained away all the land-springs overlying the London clay. As it is not uncommon for a $1\frac{1}{4}$-inch tube to yield from 500 to 600 gallons an hour, constant pumping, contractors in erecting large buildings find them very useful for supplying their engines, or for mixing mortar, and on the completion of the contracts they are taken up for use elsewhere.

This feature of being readily transportable from place to place renders them invaluable for railway contracts, and especially so for exploring expeditions. Another use is for testing ground to ascertain how deep water lies below the surface, to test the quality and quantity obtainable before sinking larger permanent tube wells. It will readily be seen how valuable such a rapid means of obtaining such data becomes, as before purchasing land for any purpose the primary question of water supply can thus be settled.

The tube, being very small, is in itself capable of containing only a limited supply of water, which would be exhausted by a few strokes of the pump; the condition, therefore, upon which these tube wells, in their most primitive form, can be effective is that there shall be a free flow of water from the outside through the apertures into the lower end of the tube.

H 2

When the stratum in which the water is found is very porous, as in the case of gravel and some sorts of chalk, the water flows freely; and a yield has been obtained in such situations as great and rapid as the pump has been able to lift, that is, 600 gallons an hour. In some other soils, such as sandy loam, the yield in itself may not be sufficiently rapid to supply the pump; in such cases, the effect of constant pumping is to draw up with the water from the bottom a good deal of clay and sand, and so gradually to form a reservoir, as it were, around the foot of the tube, in which water accumulates when the pump is not in action, as is the case in a common well. In dense clays, however, of a close and very tenacious character, the American tube well is not applicable, as the small perforations become sealed, and water will not enter the tube. When the stratum reached by driving is a quicksand, the quantity of sand drawn up from the water will be so great that a considerable amount will have to be pumped before the water will come up clear.

When rock, stone, or incompressible clay is met with, a tube cannot be driven through it without first making a hole, and removing the cores. In some cases, however, there may be many feet of loose earth which can be easily driven through; this, especially if gravel has to be passed, is a tedious process. The tubes, therefore, may be fitted with a temporary hard wooden point which will allow them to be driven through the soft earth, and when an obstruction that cannot be penetrated is met, the point is knocked out, and being wood, and in sections, it floats to the surface of the water, and leaves an open-ended tube, through which ordinary boring tools can be passed to chisel and break up the rock. A tube can frequently be driven through gravel and clay to a depth of 70 feet in a single day. To bore to the same depth in similar stratum frequently takes ten days or a fortnight. The saving that may be effected by driving through the loose stratum can, therefore, be readily appreciated, and, what is still more important, the upper part of the tubes are fixed more tightly in the ground than if a boring had been made to receive them. In some cases, however, hard strata come right to the surface, and the boring operations consequently cannot be deferred. When this is the case, instead of using a pointed tube, an open-ended steel-shod pipe is driven into the hole as the boring proceeds. As the tools pass down the pipe they do not cut so large a hole as the outside circumference, and some little trimming down of the sides is left for the steel shoe to perform.

In great depths the single tier of pipes, with which the work commenced, cannot be forced the whole way. Tubes, therefore of smaller diameter are inserted; but as, to pump by the tube-well method, air-tight joints are absolutely necessary, the final tube is continuous from the deep spring to the surface. In this way tube wells 300 or 400 feet in length are put down, and if the spring, when tapped, rises to the surface, or within, say, 25 feet of it, only an ordinary lift-pump is required to obtain the supply. Where the water does not rise to the required height, a deep-well pump can be lowered into the tube well, and worked by rods from the surface. Bored tube wells are frequently put down in sets, and connected by horizontal mains, where large supplies are required.

The system has received much development in the hands of Messrs. Le Grand and Sutcliffe, of London, and, instead of being serviceable for short depths only, tube wells have been sunk 116 feet by simple driving, and by a combination of the system with ordinary boring-tools to much greater depths with success.

CHAPTER VII.

WELL BORING AT GREAT DEPTHS.

The first well that was executed of great depth, and which gave rise to the adoption of tools which directed public attention to the art of well boring, was that for the city of Paris by Mulot, at the Abattoir of Grenelle. This was commenced in the year 1832; and after more than eight years' incessant labour, water rose, on the 26th of February, 1842, from the total depth of 1798 feet. Subsequent to this, many wells have been sunk on the Continent, with the hope of attaining the brine springs so often met with in the Rhine provinces, or the springs destined for the supply of towns, and which are even deeper than the well of Grenelle, reaching in some cases to the extraordinary depth of 2800 feet; but all of them, like the Grenelle well, of small diameter. In their construction, however, the German engineers introduced some important modifications of the tools employed; and, amongst other inventions, Euyenhausen imparted a sliding movement to the striking part of the tool used for comminuting the rock, so as to fall always through a certain distance; and thus, while he produced a uniform action upon the rock at the bottom, he avoided the jar of the tools. Kind also began to apply his system to the working of the large excavations for the purpose of winning coal. Whilst the art was in this state, and when he had already executed some very important works in Germany, Belgium, the North of France, Creuzot, and Seraing, the Municipal Council of Paris determined to entrust him with the execution of a new well they were about to sink at Passy.

In sinking the well of Passy, the weight of the trepan for comminuting the rock was about 1 ton 16 cwt., 1800 kilog.: the height through which it fell was about 60 centimètres; and its diameter was 3 feet $3\frac{7}{16}$ inches, 1 mètre. The rods were of oak, about 8 inches on the side, and the dimensions of the cutting tool were limited to 3 feet $3\frac{7}{16}$ inches because it worked the whole time in water; but generally the class of borings Kind undertook were of such a description as justified resorting to tools of great dimensions. When sinking the shafts for winning coal, his operations required to be carried on with the full diameters of 10 feet or 14 feet; and he then drove a boring of 3 feet 4 inches diameter in the first instance, and subsequently enlarged this excavation. There can be no objection to executing Artesian borings of this diameter other than the probable exhaustion of the supply; particularly as it is now known that the yield of water by these methods is proportionate to the diameter of the column; though, strange as it may appear, the first opposition to Kind's plan of sinking the well of Passy was founded upon the assumption that he would not meet with a larger supply of water from the subcretaceous formations than had been met with at Grenelle, where the diameter of the boring was at the bottom not more than 8 inches. It is now, however, proved that there is a direct gain in adopting the larger borings, not only as regards the quantity of water to be derived from them, but also in their execution, arising from the fact that the tools can be made more secure against the effects of torsion or of concussion against the sides of the excavation, which is the cause of the most serious accidents met with in well sinking.

The trepan of M. Kind contains some peculiar details, which are shown in Figs. 105, 106. The trepan is composed of two principal pieces, the frame and the arms, both of wrought-iron, with the exception of the teeth of the cutting part, which are of cast steel. The frame has at the bottom a series of holes, slightly conical, into which the teeth are inserted, and tightly wedged up, Fig. 107. These teeth are placed with their cutting edges on the longitudinal axis of the frame that receives them; and at the extremity of the frame there are formed two heads, forged out of the same piece with the body of the tool, which also carries two teeth, placed in the same direction as the others, but double their width, in order to render this part of the tool more powerful. By increasing the dimensions of these end teeth, the diameter of the boring can be augmented, so as to compensate for the diminution of the clear space caused by the tubing, necessarily introduced for security in traversing strata disposed to fall in, or for the purpose of allowing the water from below to escape at an intermediate level.

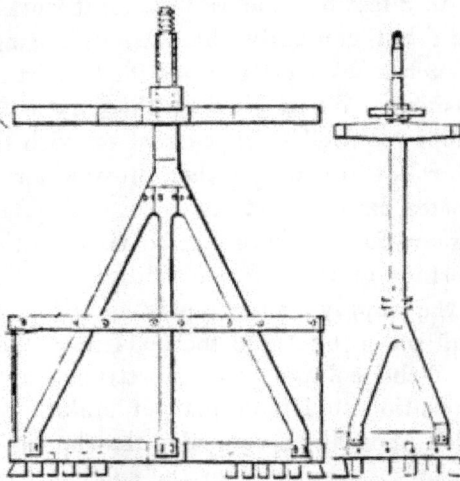

Fig. 105.   Fig. 106.

Fig. 107.

Above the lower part of the frame of the trepan is a second piece composed of two parts bolted together, and made to support the lower portion of the frame. This part of the machinery also carries two teeth at its extremities, which serve to guide the tool in its descent, and to work off the asperities left by the lower portion of the trepan. Above this, again, are the guides of the machinery, properly speaking, consisting of two pieces of wrought iron, arranged in the form of a cross, with the ends turned up, so as to preserve the machinery perfectly vertical in its movements, by pressing against the sides of the boring already executed. These pieces are independent of the blades of the trepan, and may be moved closer to it or farther away from it, as may be desired. The stem and the arms are terminated by a single piece of wrought iron, which is joined to the frame with a kind of saddle-joint, and is kept in its place by means of keys and wedges. The whole of the trepan is finally jointed to the great rods that communicate the motion from the surface, by means of a screw-coupling, formed below the part of the tool which bears the joint; this arrangement permits the

free fall of the cutting part, and unites the top of the arms and
frame, and the rod, Fig. 108. It has been proposed to substitute
for this screw-coupling a keyed joint, in order to avoid
the inconvenience frequently found to attend the rusting
of the screw, which often interposes great difficulties in
cases where it becomes necessary to withdraw the trepan.

The sliding joint is the part of Euyenhausen's invention most unhesitatingly adopted by Kind, and it is one
of the peculiarities of his system as contrasted with the
processes formerly in use. So long as his operations
were confined to the small dimensions usually adopted for
Artesian borings, he contented himself with making a
description of joint with a free fall; a simple movement
of disengagement regulating the height fixed by the machinery itself, like the fall of the monkey in a pile-driving
machine; but it was found that this system did not answer
when applied to large borings, and it also presented certain dangers. Kind then, for the larger class of borings,
availed himself of sliding guides, so contrived as to be equally
thrown out of gear when the machinery had come to the end of
the stroke, and maintained in their respective positions by being
made in two pieces, of which the inner one worked upon slides,
moving freely in the piece that communicated the motion to the
striking part of the machinery. The two parts of the tool were
connected with pins, and with a sliding joint, which, in the
Passy well, was thrown out of gear by the reaction of the
column of water above the tool unloosing the click that upheld
the lower part of the trepan, Figs. 110 to 112. The changes
thus made in the usual way of releasing the tool, and in guiding
it in its fall were, however, matters of detail; they involved no
new principle in the manner of well boring: and the modern
authorities upon the subject consider that there was something
deficient in Kind's system of making the column of water act
upon a disc by which the click was set in motion. This system, in fact, required the presence of a column of water not
always to be commanded, especially when the borings had to
be executed in the carboniferous strata.

The rods used for the suspension of the trepan, and for the transmission of the blows to it, were of oak; and this alone would constitute one of the most characteristic differences between the system of tools introduced by Kind and those made by the majority of well borers, but which, like the disengagement of the tool intended to comminute the rock, depended for its success upon the boring being filled with water. The resistance that the wood offers, by its elasticity, to the effects of any sudden jar, is also to be taken into account in the comparison of the latter with iron, for the iron is liable to change its form under the influence of this cause. The resistance to an effort of torsion need not, however, be much dwelt on, for the turn given to the trepan is always made when the tool is lifted up from its bed. For the purpose of making the rods, Kind recommended that straight-grown trees, of the requisite diameter, should be selected, rather than they should be made of cut timber, as there is less danger of the wood warping, and the character of the wood is more homogeneous. He generally used these trees in lengths of about 50 feet, and he connected them at the ends with wrought-iron joints, fitting one into the other, Fig. 112. The ironwork of the joints is made with a shoulder underneath the screw-coupling, to allow the rods to be suspended by the ordinary crow's foot during the operation of raising or lowering them. In the works executed at Passy there was a kind of frame erected over the centre of the boring, of sufficient height to allow of the rods being withdrawn in two lengths at a time, thus producing a considerable economy of time and labour.

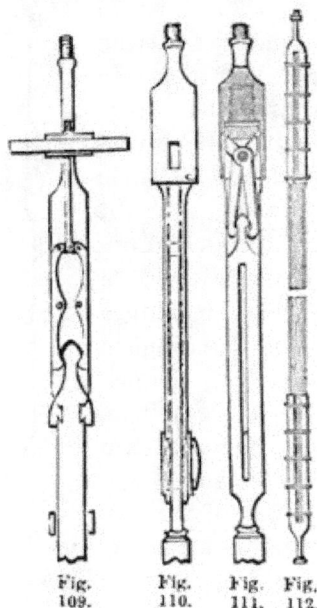

Fig. 109.   Fig. 110.   Fig. 111.   Fig. 112.

Nearly all the processes yet introduced for removing the products of the excavation must be considered to be, more or less, defective, because all are established on the supposition that the comminuting tool must be withdrawn, in order that the shell, or other tool intended to remove the products of the working of the comminutor, may be inserted. This remark applies to Kind's operations at Passy and elsewhere, as he removed the rock detached from the bottom of the excavation by a shell, Figs. 113, 114, which was a modification of the tool he invariably employs for this purpose. It consisted of a cylinder of wrought iron, suspended from the rods by a frame, and fastened to it, a little below the centre of gravity, so that the operation of upsetting it, when loaded, could be easily performed. This cylinder was lowered to the level of the last workings of the trepan, and the materials already detached by that instrument were forced into the tool, by the gradual movement of the latter in a vertical direction. Some other implements, employed by Kind for the purpose of removing the products of the excavation in the shafts for the coal mines of the North of France, were ingenious and well adapted to the large dimensions of the shafts; but they were

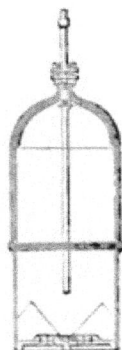

Fig. 113.

Fig. 114.

all, in some degree, exposed to the danger of becoming fixed, if used in the small borings of Artesian wells, by the minute particles of rocks falling down between their sides and the excavation from above. Their use was therefore abandoned, and the well of Passy was cleared out with the shell, the bottom of which was made to open upwards, with a hinged flap, which admitted the finer materials detached by the trepan. There were also several tools for the purpose of withdrawing the broken parts of the machinery from the excavation, or whatever substances might fall in from above; and all were marked by a great degree of simplicity, but they did not differ enough from those generally used for the same purpose to merit further remarks. But there is no doubt that Kind deprived himself of a valuable appliance in not using the ball-clack, that other well borers employ, Fig. 115.

Fig. 115.

At Passy great strength was given to the head of the striking tool, and to the part of the machinery applied to turn the trepan, because the great weight of the latter superinduced the danger of its breaking off under the influence of the shock, and because the solidity of this part of the machinery necessarily regulated the whole working of the tool. The head of the boring arrangement was connected with the balance-beam of the steam-engine by a straight link-chain, with a screw-coupling, admitting of being lengthened as the trepan descended, Figs. 116, 117. The balance-beam, in order to increase its elastic force in the upward stroke, is in Kind's works made of wood, in two pieces; the upper one being of fir and the lower one of beech. The whole of the machinery is put in motion by steam, which is admitted

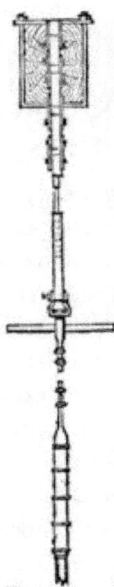

Fig. 116.   Fig. 117.

to the upper part of the cylinder, and presses it down, and thus raises the tool at the other end of the beam to that part in connection with the cylinder. The counterpoise to the weight of the tools is also placed upon the cylinder-end of the beam. The cylinder receives the steam through ports that are opened and closed by hand, like those of a steam-hammer; so that the number of the strokes of the piston may be increased or diminished, and the length of the strokes may be increased, as occasion may require.

The balance-beam is continued beyond the point where the piston is connected with it, and it goes to meet the blocks placed to check the force of the blow given by the descent of the tool. The guides of the piston-head are attached to the part of the machinery that acts in this manner; but at Passy, Kind made the balance-beam work upon two free plummer-

blocks, or blocks having no permanent cover, that they might
be more easily moved whenever it was necessary to displace the
beam, for the purpose of taking up or letting down the rods, or
for changing the tools; for the balance-beam was always immediately over the centre of the tools, and it therefore had to be
displaced every time that the latter were required to be changed.
This was effected by allowing the beam to slide horizontally, so
as to leave the mouth of the pit open. The counter-check,
above mentioned, likewise prevented the piston from striking
the cylinder cover with too great a force, when it was brought
back by the weight of the tools to its original position. The
operation of raising and lowering the rods, or of changing the
tools, was performed at Passy by a separate steam-engine, and
the shell was discharged into a special truck, moving upon a
railway expressly laid for this purpose in the great tower
erected over the excavation. All these arrangements were in
fact made with the extreme attention to the details of the
various parts of the work which characterises the proceedings
of continental engineers, and conduces so much to their success.

The beating, or comminution of the rock, was usually effected
at Passy at the rate of from fifteen strokes to twenty strokes a
minute. The rate of descent, of course, differed in a marked
manner according to the nature of the rock operated upon; but,
generally speaking, the trepan was worked for the space of
about eight hours at a time, after which it was withdrawn, and
the shell let down in order to remove the *débris*. The average
number of men employed in the gang, besides the foreman, or
the superintendent of the well, was about fourteen: they consisted of a smith and hammerman, whose duty it was to keep the
tools in order; and two shifts of men entrusted with the excavation, namely, an engine-driver and stoker, a chief workman
or sub-foreman, and three assistants. The total time employed
in sinking the shafts executed upon this system in the North of
France, where it has been applied without meeting with the
accidents encountered in the Passy well, was found to be suscep-
tible of being divided in the following manner: from 25 per
cent. to 56 per cent. was employed in manœuvring the trepan;

from 11 per cent. to 14½ per cent. in raising and lowering the
tools; from 19 per cent. to 21 per cent. in removing the materials detached from the rocks, and cleaning out the bottom of
the excavation; and from 8 per cent. to 10½ per cent. was lost,
owing to the stoppage of the engines, or to the accidents from
broken tools, or to other causes always attending these operations.   In the well of Passy there was, of course, a considerable
difference in the proportions of the time employed in the various
details of the work : and the long period occupied in obviating
the effects of the slips which took place in the clays, both in
the basement beds of the Paris basin and in the subcretaceous
strata, would render any comparison derived from that well of
little value; but it would appear that, until the great accident
occurred, the various operations went on precisely as Kind had
calculated upon.

Kind-Chaudron System.

In the year 1872 Emerson Bainbridge, C.E., drew attention
to the Kind-Chaudron system of sinking mine shafts through
water-bearing strata, without the use of pumping machinery, in
a paper read before the Institute of Civil Engineers. As the
operation is almost identical with that which would have to be
carried through in the case of a well sunk through an upper
series of water-bearing strata, of minor importance or of impure
quality, past rock and into the lower water strata, as for instance
through tertiaries and chalk into the lower greensand, the
following extract from Bainbridge's paper may be read with
interest.

In the first place, it may be desirable to describe briefly the
system of sinking hitherto pursued in passing through strata
yielding large quantities of water. The most important sinkings of this character have been carried out in the county of
Durham, to the east of the point at which the Permian overlie
the carboniferous rocks. In this district there is a thin bed of
sand between the Permian rock and the coal measures. Towards
this bed the feeders of water are generally found to increase,
and in the sand there is usually a large reservoir of water. The

mode of sinking will be understood by reference to Fig. 118. Whilst sinking in hard rock, it has ordinarily been the custom to place iron curbs, or cribs, wherever a bed of stone appeared to form a natural barrier between two distinct feeders of water. Thus it has frequently happened that important feeders have been tubbed back, rendering much less pumping power necessary than would have been required had all the feeders been allowed to accumulate in the shaft. As will be seen by Fig. 118, the number of wedging cribs employed is no less than thirteen in 250 feet. The cribs forming the foundation of each set of tubbing are generally much more massive and costly than the segments of tubbing.

The process of fixing the crib is as follows;— The diameter of the shaft is made about 30 inches larger than that of the inside of the tubbing. When a bed of rock, which may be considered

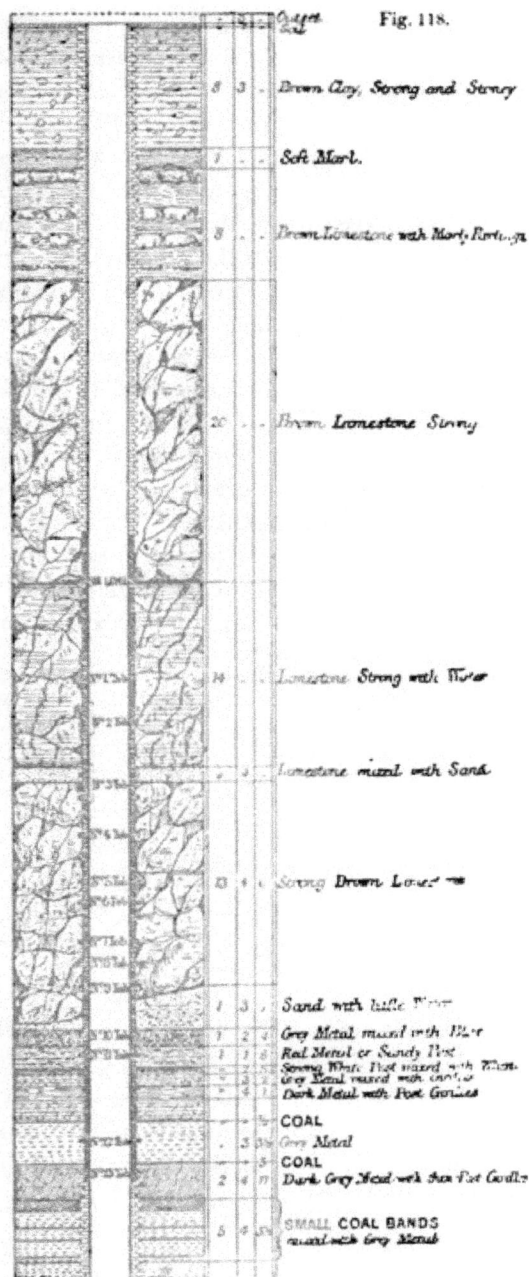

Fig. 118.

sufficiently hard and close to separate the feeders above and
below it, is reached, the shaft is contracted to the diameter
of the tubbing, and a smooth horizontal face is made on which
to place the wedging crib. The wedging crib, which usually
consists of segments of about 4 feet long by 6 inches high by
14 inches wide, is then placed on the bed. To give the crib a
firm and secure position, it is tightly wedged with wood, both
behind and between the joints; the tubbing is then built upon it
to the next wedging crib, which rests upon a bell-shaped section
of rock. When the tubbing nearly reaches this crib, the rock is
removed piece by piece, and the top ring of tubbing is placed
close up against the crib. It will thus be seen that the fixing of
each crib is a costly process, often causing considerable delay.

In some cases, where it has been difficult to find suitable
foundations for intermediate wedging cribs, the whole of the
water-bearing rocks have been sunk through without attempting
to stop the feeders separately, and no tubbing has been placed
in the shaft till the wedging crib could be fixed below the lowest
feeder. This process is more expeditious where there are small

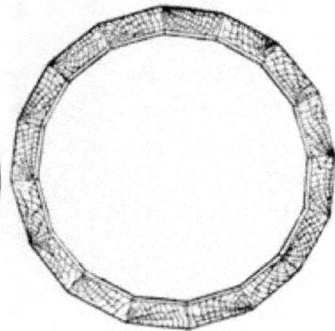

Fig. 119.   Fig. 120.

quantities of water; but where the water is excessive greater
delay is caused by contending with it than from putting in
numerous sets of tubbing to stop the feeders separately. The
tubbing used in England has almost invariably been of cast
iron; on the Continent, till recently, tubbing of wood has chiefly
been used. Illustrations of both descriptions are shown by
Figs. 119 and 120.

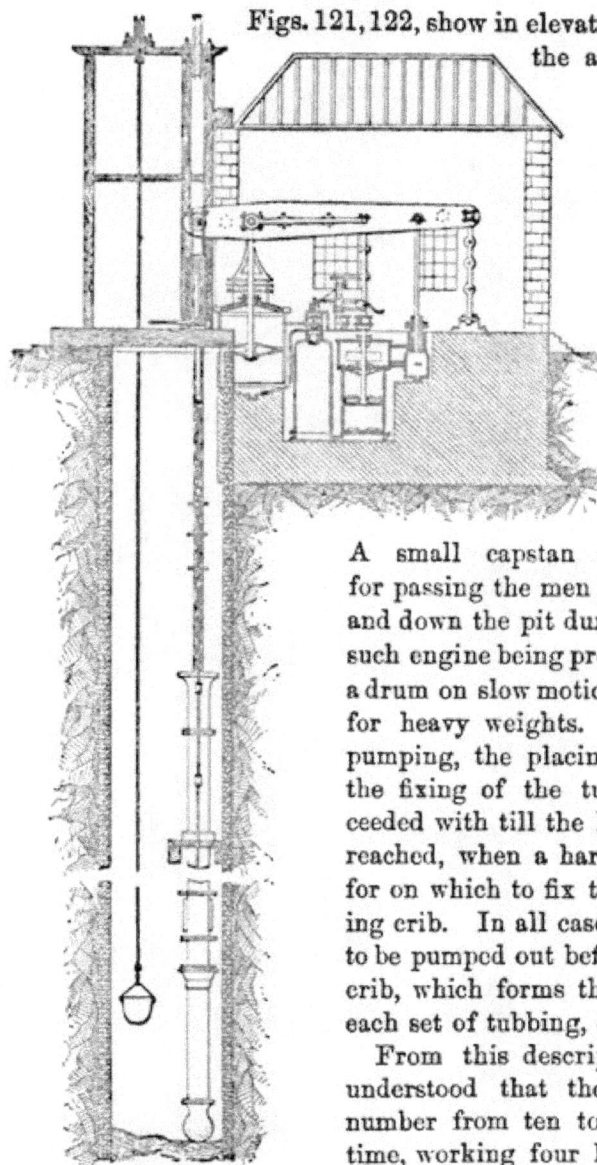

Fig. 121.

Figs. 121, 122, show in elevation the plant and the arrangements generally in use at extensive sinkings. Where the water is in large quantities it is usually pumped by an engine erected for the purpose, assisted by the engine or engines intended to be employed to raise the coal. A small capstan engine is used for passing the men and material up and down the pit during the sinking, such engine being provided also with a drum on slow motion, which is used for heavy weights. The continual pumping, the placing of cribs, and the fixing of the tubbing are proceeded with till the lowest feeder is reached, when a hard bed is sought for on which to fix the lowest wedging crib. In all cases the water has to be pumped out before the wedging crib, which forms the foundation of each set of tubbing, can be placed.

From this description it will be understood that the sinkers, who number from ten to twelve at one time, working four hours at a shift in a pit, say, 14 feet in diameter, are compelled to work in water until all the tubbing is fixed.

I

This causes a serious obstacle to blasting, and in other ways delays the progress of the work.

The tubbing used for damming back the water is generally in segments from 1 foot to 3 feet high, and about 4 feet in length, the thickness

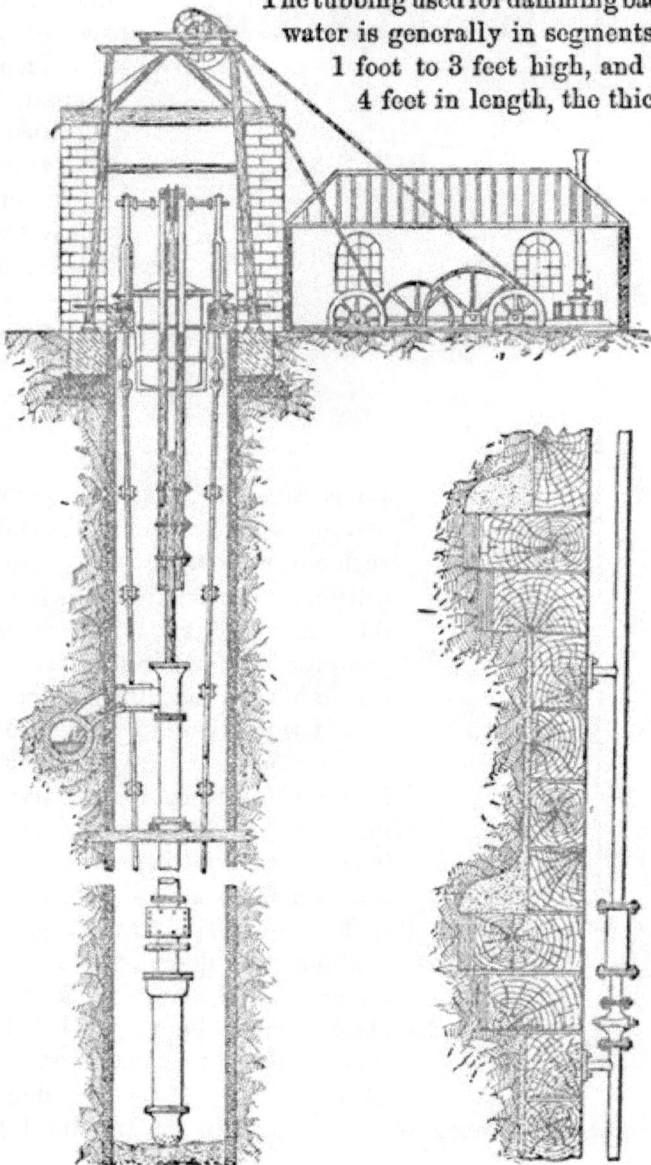

Fig. 12

Fig. 123.

varying from half an inch to $3\frac{3}{4}$ inches. It is kept in position by packing with wood behind the joints; and is made water-tight by placing between the segments pieces of wood sheeting about half an inch thick, which are wedged when all the tubbing is fixed, usually twice with wood, and sometimes once with iron wedges.

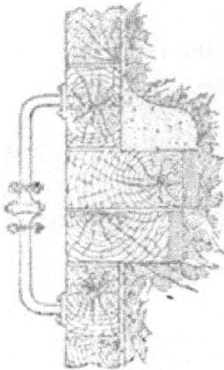

Fig. 124.

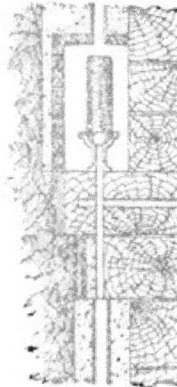

Fig. 126.

To equalise the pressure of water and gas behind the different sets of tubbing, pass pipes, Figs. 123 and 124, are sometimes used. Another expedient to effect this is to have a valve, working upwards, placed in the wedging crib, Fig. 125. A ball is also sometimes used, Fig. 126.

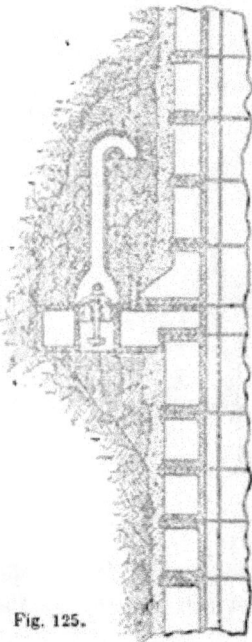

Fig. 125.

The various modes of piercing beds of quicksand are;—By hanging tubbing to that already fixed, and adding fresh rings as the sand is removed. This is only practicable when the quantity of sand is inconsiderable. By heavily weighting a cylinder of iron of the same size as the shaft, and thus forcing it down through the sand. By keeping back the sand by the use of piles—a resource that can only be recommended when the bed of sand is not of great thickness. When the water is excessive, by using pneumatic agency. As these operations are apart from our immediate subject we need not further discuss them.

M. Chaudron's system, which is a

modification of Kind's, is divisible into the following distinct processes, which consist of;—

The erection of the necessary machinery on the surface, and the opening of the mine.

The boring of the pits to the lowest part of the water-bearing strata.

The placing of the tubbing.

The introduction of cement behind the tubbing, to complete its solidity.

The extraction of water from the pits, and the placing of the wedging cribs, or "faux cuvelage," below the moss box.

Figs. 127 to 129 show in elevations and in plan the plant usually employed on the surface. O is a small capstan engine, having a cylinder 20 inches in diameter and a stroke of 32 inches, working on the third motion. Attached to this engine, and working in the small pit C, is a counterbalance weight.

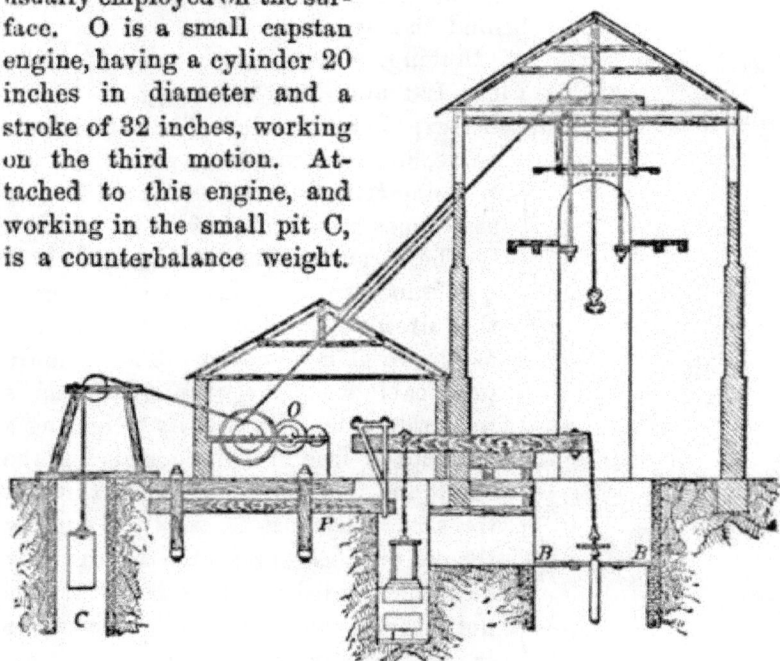

Fig. 127.

This engine is used for raising and lowering boring tools, and for lifting the *débris* resulting from the boring. As far as the platform, which is about 10 feet from the surface, the pit has a diameter of 19 feet, or 4 feet more than the diameter of the pit

below. At a level of about 38 feet above this platform there is
a tramway on which small trucks run, carrying the *débris*
cylinder on one side, and the boring tools at the other. At a
level of 48 feet above the platform are placed supports for the
wooden spears to which the boring tools are attached. The
machinery for boring is worked by a cylinder, which has a dia-
meter of $39\frac{1}{3}$ inches, and a full stroke of $39\frac{1}{3}$ inches, the usual
stroke varying from 2 feet to 3 feet. A massive beam of wood
transmits motion from this cylinder to the boring apparatus, the
connection between the beam and the piston-rod and the beam
and the boring tools being made by a chain. The engine-man
sits close to the engine, and applies the steam above the piston
only. The down stroke of the boring tools is caused by the

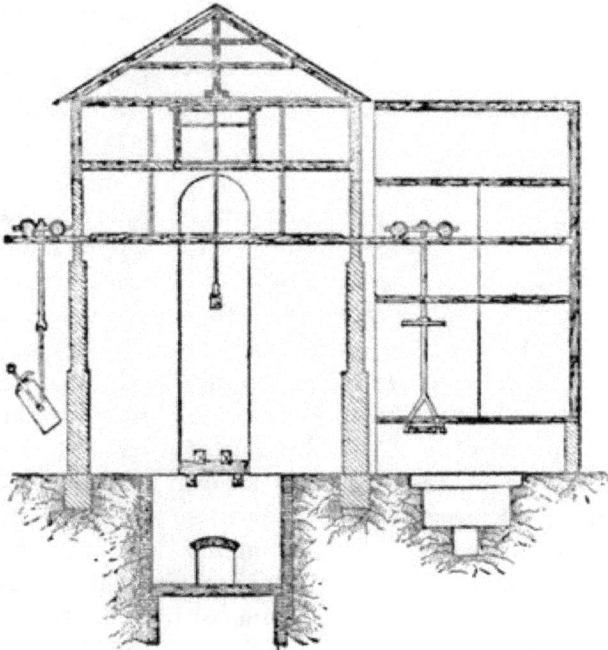

Fig. 123.

sudden opening of the exhaust, and a frame then prevents the
shock of the boring rods from being too severe. The engines
work at speeds varying from 12 to 18 strokes a minute, according to the character of the strata passed through.

After the working platform is fixed, the first boring tool applied is the small trepan, Figs. 130 to 133. This tool is attached to the wooden beam by the same arrangement shown

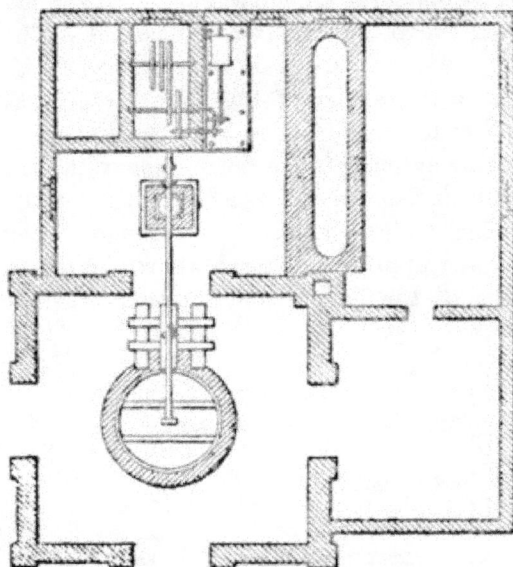

Fig. 129.

by Fig. 117. The boring tools can be lowered at pleasure by means of an adjusting screw. Next in order comes the handle for boring. This is worked by four men on the platform, and is turned by the aid of a swivel. Attached to the handle-piece are wooden rods, made from Riga pitch pine. These rods are 59 feet in length and $7\frac{3}{4}$ inches square. A swivelled

ring, Figs. 134, 135, is attached to the rope when raising and lowering the boring rods. The small trepan cuts a hole 4 feet $8\frac{3}{4}$ inches in diameter, and has fourteen teeth, fitted in cylindrical holes and secured by pins entering through circular slots. The teeth are steeled. At a distance of 4 feet 4 inches above the main teeth of the trepan is an arm B, with a tooth at each end. This piece answers the purpose of a guide, and at the same time removes irregularities from the sides of the hole. At a distance of 13 feet 6 inches above the main teeth are the actual guides, consisting of two strong arms of iron fixed on the tool, and placed at right angles to each other. The hole made by the small trepan is not kept at any fixed distance in advance of the full-sized pit, but the distance generally varies from 10 to 30 yards. With the small trepan, which weighs 8 tons, the progress varies from 6 to 10 feet a day.

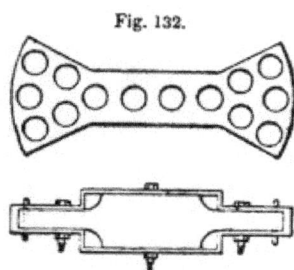

Fig. 132.

Plan of Guide B.
Fig. 133.

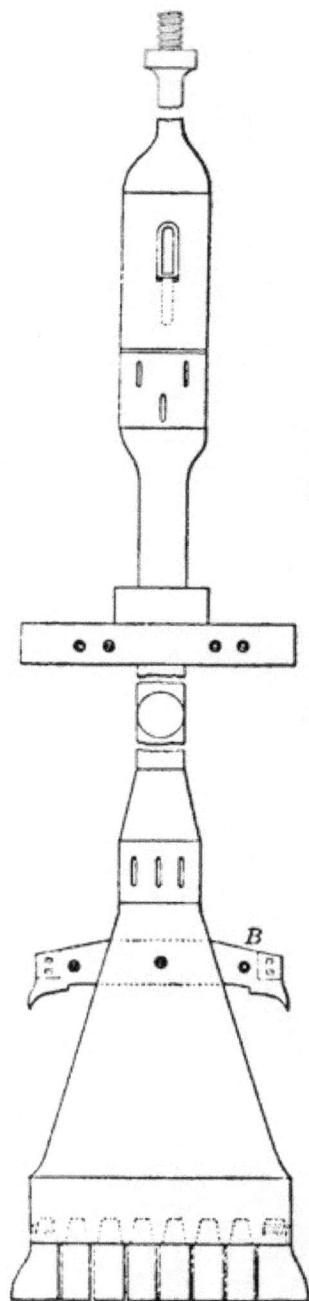

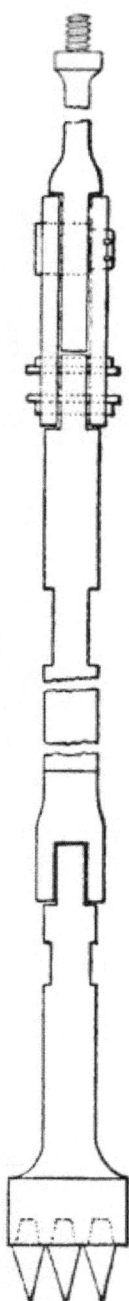

The large trepan, Figs. 136 to 138, weighs $16\frac{1}{2}$ tons, is forged in one solid piece, and has twenty-eight teeth. A projection of iron forms the centre of this trepan and fits loosely into the hole made by the small trepan, acting as a guide for the tool. At a distance of 7 feet 6 inches above the teeth, a guide is sometimes fixed on the frame, but is not furnished with teeth. At a distance of 13 feet

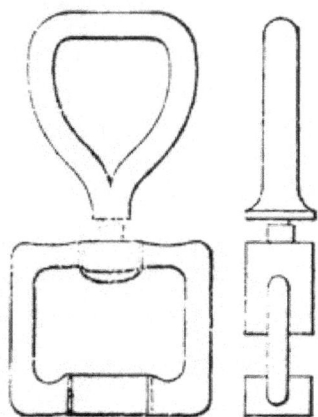

Fig. 130.

FIG. 131.

Fig. 134.

Fig. 135.

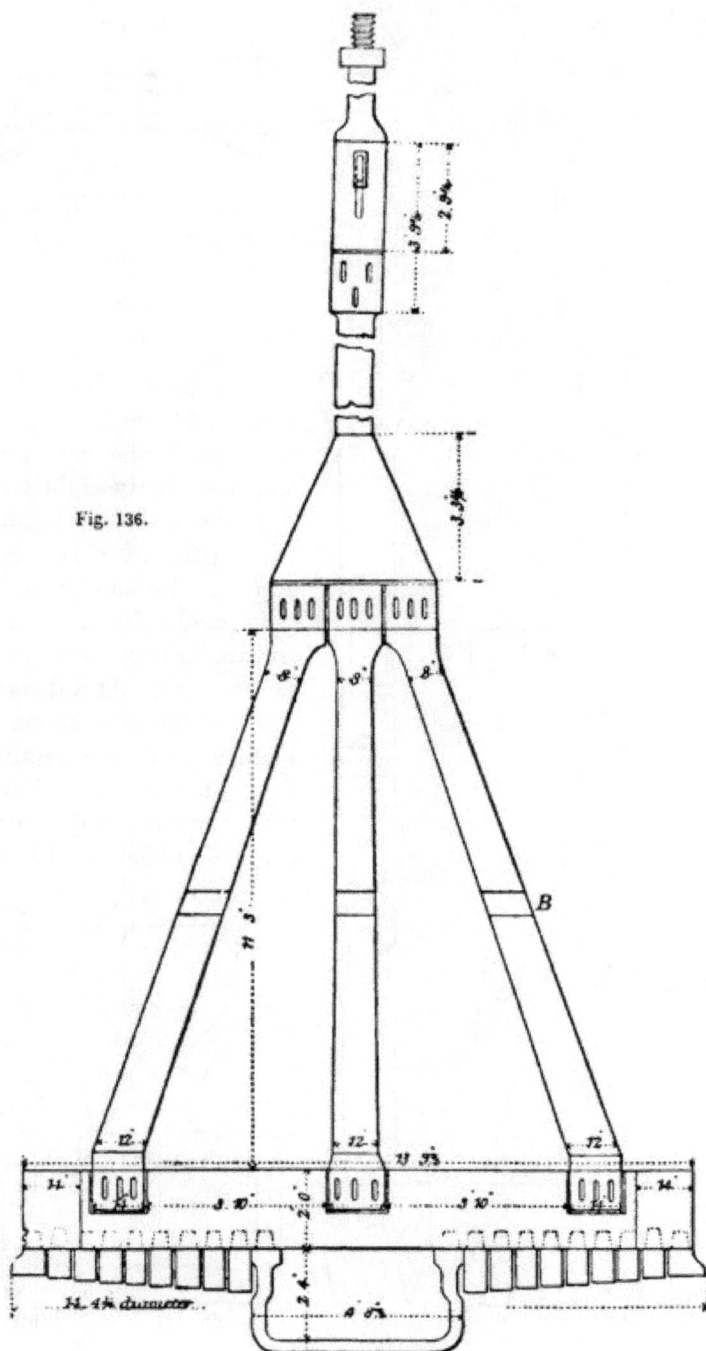

Fig. 136.

3 inches from the teeth are two other guides at right angles to
each other. These guides are let down the pit with the boring
tool, the hinged part of the guides being raised whilst passing

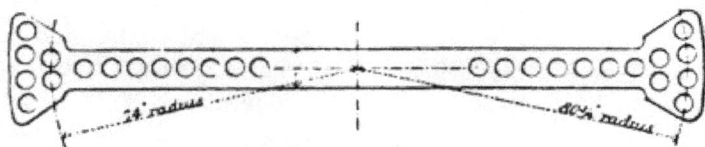

Fig. 138.

through the beams at the top of the pit, which are only 6 feet
7 inches apart. When the tool is ready to work, the two arms
are let down against the side of the pit, and are hung in the
shaft by ropes, thus acting as a guide
for the trepan, which moves through
them. To provide against a shock to
the spears when the trepan strikes the
rock on the down stroke, at the upper
part of the frame a slot motion is arranged, the play of which amounts to
about half an inch. The teeth of the large
trepan are not horizontal, but are deeper
towards the inside of the pit, the face
of the inside tooth being $3\frac{3}{4}$ inches lower
than the outside. The object of this is
to cause the *débris* to drop at once into
the small hole, by the face of the rock
at the bottom of the pit being somewhat inclined. The teeth used, Figs.
139 to 142, are the same both for the
large and the small trepan, and weigh
about 72 lb. each. As a rule, only one
set of teeth is kept in use, this set

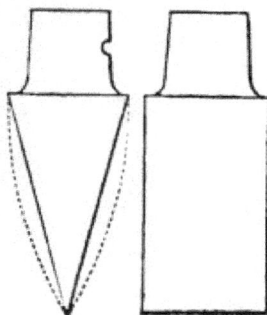

Fig. 139.　　Fig. 140.

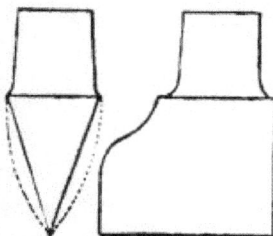

Fig. 141.　　Fig. 142.

working for twelve hours, the alternate twelve hours being
employed in raising the *débris*. This time is divided in about
the following proportions:—Boring, twelve hours ; drawing the
rods, one hour to five hours, according to depth ; raising the
débris, two hours ; and lowering the rods, one hour to five hours.

The maximum speed of the larger trepan may be taken at about 3 feet a day. The ordinary distance sunk is not more than 2 feet a day, and in flint and other hard rocks the boring has proceeded as slowly as 3 inches a day.

The *débris* in the small bore-hole contains pieces of a maximum size of about 8 cubic inches. In the large boring, pieces of rock measuring 32 cubic inches have been found. As a rule, however, the material is beaten very fine, having much the appearance of mud or sand. In both the large and the small borings the *débris* is raised by a shell, similar to Figs. 113, 114, and in this system consisting of a wrought-iron cylinder, 3 feet 3 inches in diameter by 6 feet 9 inches long, and containing two flap valves at the bottom, through which the excavated material enters. This apparatus is passed down the shaft by the bore rods, and it is moved up and down through

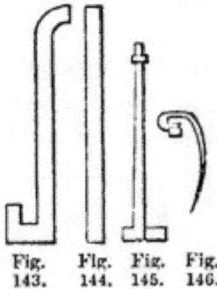

Fig. 143. Fig. 144. Fig. 145. Fig. 146.

Fig. 147. Fig. 148.

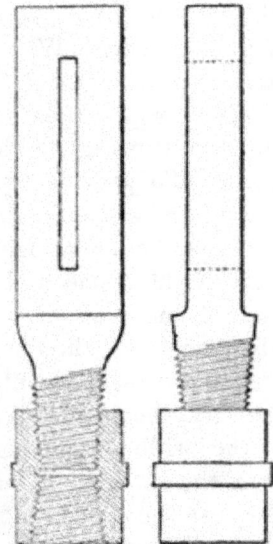

Fig. 149. Fig. 150.

a distance varying from 6 to 8 inches, for about a quarter of an hour, and is then drawn up and emptied. In some cases where the rock is hard, three sizes of trepan are used consecutively, the sizes being 5 feet, 8 feet, and 13 feet.

The several other tools and appliances used during the boring operations are shown, Figs. 143 to 148, including the key, Figs. 147, 148, used at the surface to disconnect the rods, the hook on which each rod is hung after being raised to the high platform and there detached, the bar upon which the hooks are moved,

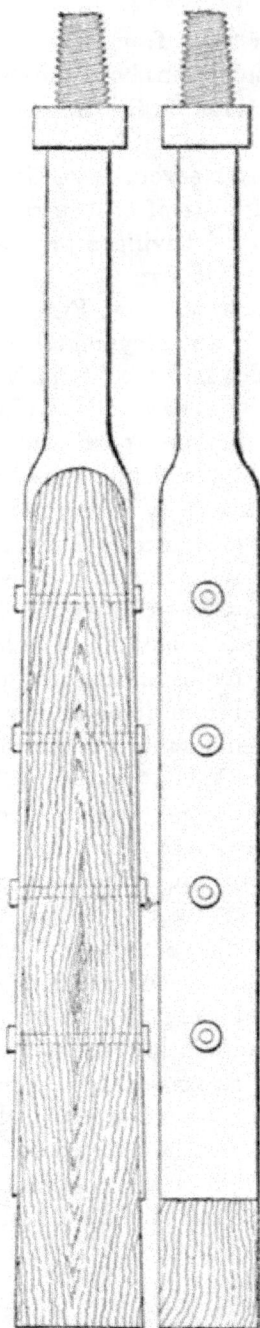

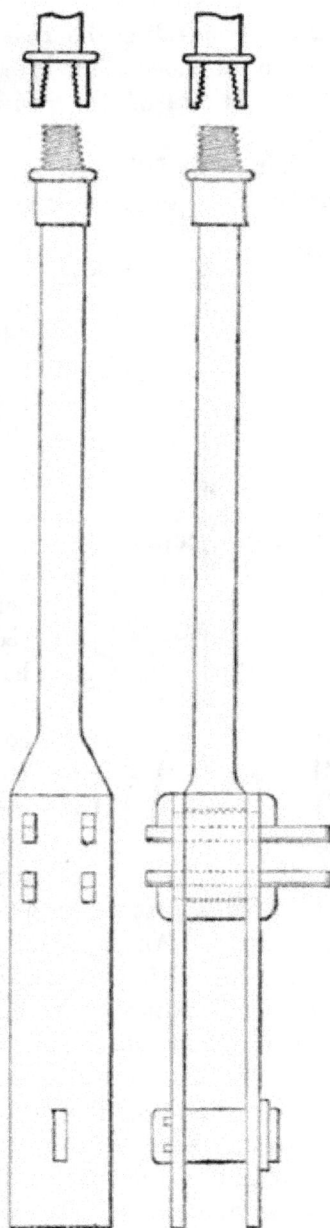

Fig. 151.

Fig. 152.

Fig. 153.

Fig. 154.

and the fork for suspending the rods or tools from the rollers when it is desired to move the rods or tools from above the shaft.

Figs. 149 to 154 are of the connections to the trepan and spears or rods.

Should broken tools fall into the shaft, several varieties of apparatus are used for their recovery. In case of broken rods of any kind having a protuberance that can be clutched, a hook or crow, Figs. 145, 146, of an epicycloidal form, enables the object to be taken hold of very readily. Where the broken part has no shoulder which can be held, but is simply a bar, the apparatus shown by Figs. 155, 156, is employed. This is composed of two parts. The rods, the bottom of which have teeth inside, are prevented from diverging by the cone and slide on the main rods. When passed over a rod or pipe, they clutch it by means of the teeth, and draw it up. Chaudron has, by this tool, raised a column of pipes 295 feet in length and 8 inches in diameter. An instrument, called a "grapin," Figs. 157, 158, is used for raising broken teeth or other small objects which may have fallen into the bottom of the shaft. This tool also has one part sliding in the other, and is lowered with the claws closed. The parts are moved by two ropes worked from the surface. By weighting the cross-bar, which is attached to the moving parts, the pressure desired can be exerted on the claws. The weight is then lifted, the claws are opened, and are made to close upon the substance to be raised. This instrument is now seldom required.

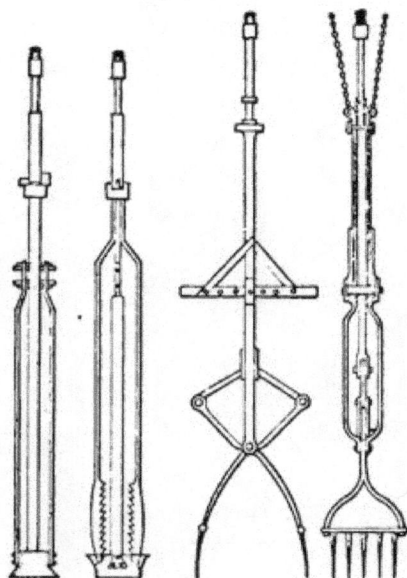

Fig. 155. Fig. 156.   Fig. 157.   Fig. 158.

In boring shafts in the manner described, without being able to prove in the usual way the perpendicularity of the shaft, it might be feared that the system would be open to objection on this account. It appears, however, that in all cases where Chaudron has sunk shafts by this system, he has succeeded in making them perfectly vertical. This is ensured by the natural effect of the treble guide, which the chisels and the two sets of arms attached to the boring tools afford, and by the fact that if the least divergence from a plumb-line is made by the boring tool, the friction of the tool upon one side of the shaft is so great as to cause the borers to be unable to turn the instrument.

Boring alternately with the large and the small instrument, the shaft is at length sunk to the point at which the lowest feeder of water is encountered. In a new district this has to be taken, to some extent, at hazard ; but where pits have been sunk previously, it is not difficult to tell, by observing the strata, almost the exact point at which the bottom of the tubbing may be safely fixed. This point being ascertained, the third process is arrived at.

As the object of placing tubbing in a shaft is effectually to shut off the feeders, which for water supply may have some bad qualities, and to secure a water-tight joint at the base, it is important that the bed on which the moss box has to rest should be quite level and smooth. This is attained by the use of a tool, termed a " scraper," attached to the bore-rods, the blades being made to move round

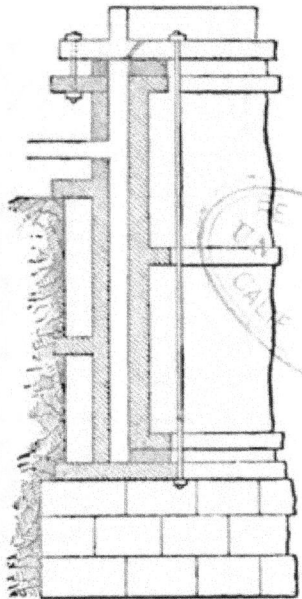

Fig. 159.

the face of the bed intended for the moss box. The tubbing employed is cast in complete cylinders. At Maurage each ring has an internal diameter of 12 feet and is 4 feet 9 inches high. Each ring has an inside flange at the top and bottom, and also

a rib in the middle, the top and bottom of the ring being turned
and faced. The rings of tubbing are attached to each other by
twenty-eight bolts 1·1 inch in diameter, passed through holes
bored in the flanges. The tubbing is suspended in the pit by
means of six rods, which are let down by capstans placed at a
distance of 30 feet above the top of the pit. These machines
work upon long screws. When a new ring of tubbing is added,
the rods are detached at a lower level, and are hung upon chains,
thus leaving an open space for passing it forward. Before each
ring is put into the pit it is tested by hydraulic apparatus, Fig.
159. The tubbing is usually proved to one-half more pressure
than it is expected to be subjected to. At Maurage, for a
length of 550 feet of tubbing, the chief particulars respecting
it are :—

	Length.	Thickness.	Pressure expected.	Pressure at which Tubbing is proved.
	feet.	inches.	lb. a square inch.	lb. a square inch.
Top	130	1·17	30	45
	60	1·31	60	90
	60	1·57	90	135
	60	1·76	120	180
	60	1·96	150	225
	60	2·16	180	270
	60	2·35	210	315
Bottom ..	60	2·55	240	360

The joints between the rings of tubbing are made with sheet
lead one-eighth of an inch thick, coated with red-lead. The lead
is allowed to obtrude from the joint one-third of an inch, and is
wedged up by a tool which has a face one-twelfth of an inch
thick. The mode of suspending the tubbing to the rods will be
understood by referring to Figs. 160 to 162. The rods are
attached to a ring by the bolts connecting one ring of tubbing
with another. The bottom ring of tubbing and the ring carrying the moss box have their top flange turned inwards, but their
bottom flange outwards. A strong web of iron, forming the base

of a tube 16½ inches in diameter, is attached to the tubbing. The object of this tube is to cause the water in the shaft to ease the suspension rods, by bearing part of the weight of the tubbing.

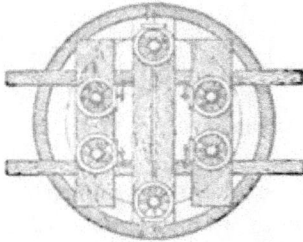

Fig. 160.

Fig. 161.

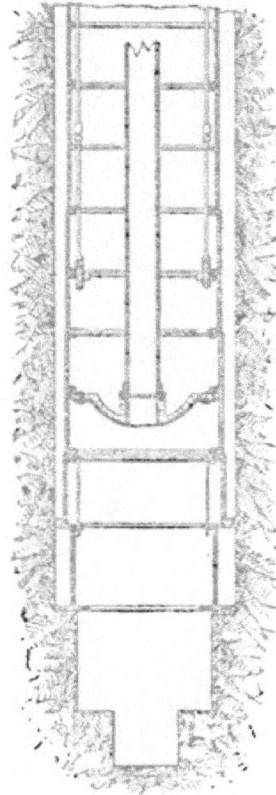

Fig. 162.

Cocks to admit water are placed at intervals up the tube, by which means the weight upon the rods can be easily regulated, so that not more than one-tenth to one-twentieth of the weight of the tubbing is suspended by the rods at one time. The ring holding the moss box is hung from the bottom joint in the tubbing by sliding rods.

The arrangement of the moss box which forms the base of the tubbing is one of the most important points requiring attention in this system of sinking. Ordinary peat moss is used. It is enclosed in a net, which, with the aid of springs, keeps it in its place during the descent of the tubbing. When the moss box, which hangs on short rods fixed to the tubbing, reaches the face of rock, it is dropped gently upon it, and the whole weight of the tubbing is allowed to rest upon the bed. This compresses the moss, the capacity of the chamber holding it is diminished, and the moss is forced against the sides of the shaft, thus forming a water-tight joint, past which no water can escape. This completes the third process.

It may be noted that up to this point the following important differences between this and the ordinary system of placing tubbing are to be observed ;—The tubbing, on reaching its bed, bears the aggregate pressure of all the feeders of water which have been met with in the shaft. The tubbing, having been passed down the shaft in the manner described, no wedging behind, or other modes of consolidating it in the shaft, have been carried out. The connection between each ring of tubbing is so carefully made that the repeated wedging of the joints, as in the ordinary system, is rendered unnecessary. The pit is still full of water up to the ordinary level.

Under these conditions the next process is ;—The introduction of cement behind the tubbing to complete its solidity.

Before the water is removed, the annular space between the tubbing and the sides of the shaft is filled with hydraulic cement, to render the tubbing impermeable, by a process of consolidation less liable to the effect of any pressure of water or gas which may be exerted towards the centre of the shaft. The cement is inserted behind the tubbing by close ladles, Figs. 163, 164, capable of holding 44 gallons, and consisting of two iron plates, one-eighth of an inch thick, fixed on two wooden uprights $3\frac{1}{8}$ inches square. This apparatus is curved to suit the mean circumference of the space to be concreted. A piston is placed at the top of the ladle, and to this piston is attached a rod, which can be moved from the surface ; a door is also attached to the piston. The ladle containing the concrete is passed down behind the tubbing by means of a windlass at the surface, and when it reaches the lowest point, the piston is pushed down and the cement allowed to escape from the chamber. The weight of the cement and the ladle is sufficient, with a little ballast, to enable it to descend easily.

A number of experiments have been made to discover a cement which will not harden too quickly, and which, when hardened, will form a perfectly compact and solid mass. A composition having the following proportions has been found the best :—Hydraulic lime, from the lias near Metz, slaked by sprinkling, 1 part; picked sand, from the Vosges sandstone, 1 part; trass, from Andernacht on the Rhine, 1 part; cement from Ropp, Haute Saone, ¼ part.

Six men are employed in putting in the cement :—two at the windlass for letting down the ladle, two for working the rods attached to the piston, and two on the working platform. The rods referred to have been found such an inconvenience that lately a rope on another windlass has been used, and an appliance arranged for dropping the piston by moving the rope.

Fig. 163.   Fig. 164.

When a sufficient time has elapsed for the centre to harden, the water within the tubbing, now effectually separated from the feeders, is drawn out by a bucket worked by the crab engine,— an operation which occupies from one to three weeks, according to circumstances. When concluded the joint between the moss box and the rock bed can be examined. In some cases this joint is considered sufficient; but it is generally thought desirable to form a base to the tubbing by building a few feet of brickwork in cement on a ring or crib of wood, as in Fig. 165. Another wooden crib is then placed on the top of this brickwork, and

K

above this, two cast-iron segmental wedging cribs with a broad

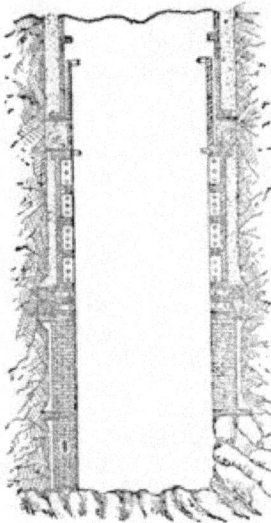

bed also wedged perfectly tight. On the base so prepared, four or more rings of tubbing in segments are fixed, the top ring coming close against the bottom of the moss box. This being done the work is completed, and the sinking of the shaft is continued in the ordinary way.

The application of the boring trepan is not to be recommended in the sinking of the dry part of the shaft. The use of the tool would cause the sinking to extend over a longer period, since the breaking of the rock passed through into such minute particles would lead to loss of time.

Fig. 165.

DRU'S SYSTEM.

The system applied by Dru is worthy of attention, not so much on account of the novelty of the invention, or of any new principle involved in it, as on account of the contrivances it contains for the application of the tool, " *à chute libre*," or the free-falling tool, to Artesian wells of large diameters. It has been already explained that under Kind's arrangements the trepan was thrown out of gear by the reaction of the water which was allowed to find its way into the column of the excavation; but that it is not always possible to command the supply of the quantity necessary for that purpose; and even when possible, the clutch Kind adopted was so shaped as to be subject to much and rapid wear. Dru, with a view to obviate both these inconveniences, made his first trepan similar to that shown in Fig. 109, in which it will be seen that the tool was gradually raised until it came in contact with the fixed part of the upper machinery, when it was thrown out of gear. The bearings of the clutch were parallel to the horizontal line, and were found in practice to be more evenly worn, so that this instrument could be worked sometimes from eight days to fourteen days without intermission; whereas, on Kind's system, the trepan was frequently withdrawn after two days' or three days' service.

We take the following complete account of the system from a paper read by M. Dru at the Conservatoire des Arts et Métiers, Paris, 6th June, 1867.

It will be seen from Figs. 166, 167, that the boring rod A is suspended from the outer end of the working beam B, which is made of timber hooped with iron, working upon a middle bearing, and is connected at the inner end to the vertical steam cylinder C, of 10 inches diameter and 39 inches stroke. The stroke of the boring rod is reduced to 22 inches, by the inner end of the beam being made longer than the outer end, serving as a partial counterbalance for the weight of the boring rod.

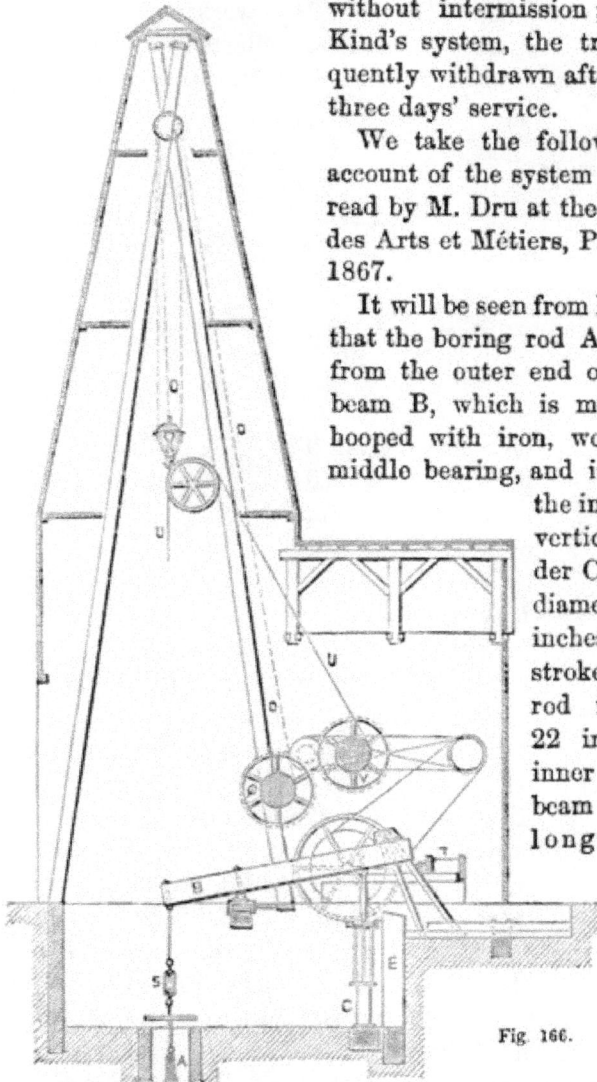

Fig. 166.

The steam cylinder is shown enlarged in Fig. 168, and is single-acting, being used only to lift the boring rod at each stroke, and the rod is lowered again by releasing the steam from the top side of the piston ; the stroke is limited by timber stops both below and above the end of the working beam B.

The boring tool is the part of most importance in the apparatus, and the one that has involved most difficulty in maturing

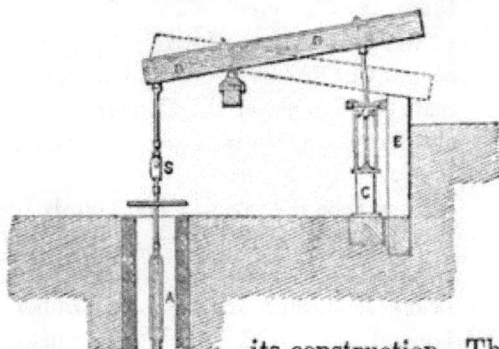

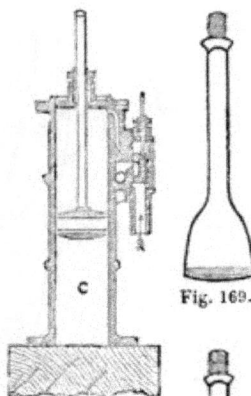

Fig. 169.

its construction. The points to be aimed at in this are,—simplicity of construction

Fig. 168.

and repairs ; the greatest force of blow possible for each unit of striking surface; and freedom from liability to get turned aside and choked.

The tool used in small borings is a single chisel, as shown in Figs. 169, 170 ; but for the large borings

Fig. 170.

it is found best to divide the tool-face into separate chisels, each of convenient size and weight for forging. All the chisels, however, are kept in a straight line, whereby

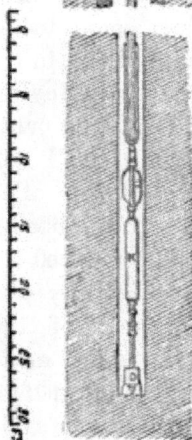

Fig. 167.

the extent of striking surface is reduced, and the tool is rendered less liable to be turned aside by meeting a hard portion of flint on a single point of the striking edge, which would diminish the effect of the blow.

The tool is shown in Figs. 171 to 177, and is composed of a wrought-iron body D, connected by a screwed end E to the boring rod, and carrying the chisels F F, fixed in separate

sockets and secured by nuts above; two or four chisels are
used, or sometimes even a greater number, according to the

Fig. 171.

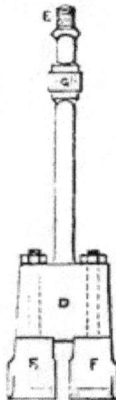

Fig. 172.

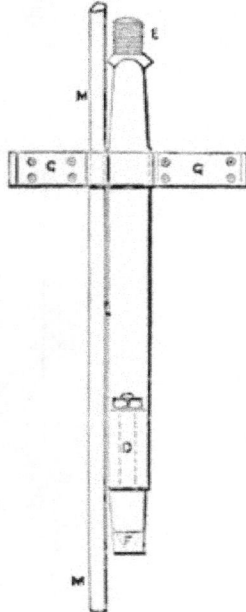

Fig. 174.

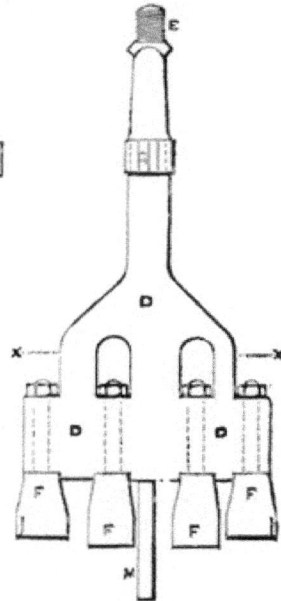

Fig. 175.

Fig. 173.

Fig. 176.

size of the hole to be bored. This
construction allows of any broken
chisel being easily replaced; and

Fig. 177.

also by changing the breadth of the two outer chisels,
the diameter of the hole bored can be regulated exactly
as may be desired. When four chisels are used, the two
centre ones are made a little longer than the others, as
shown in Fig. 175, to form a leading hole as a guide to
the boring rod. A cross-bar G, of the same width as the
tool, guides it in the hole in the direction at right angles
to the tool; and in the case of the larger and longer tools
second cross-bar higher up, at right angles to the first and
parallel to the striking edge of the tool, is also added.

If the whole length of the boring rod were allowed to fall
suddenly to the bottom of a large bore-hole at each stroke,

frequent breakages would occur; it is therefore found requisite
to arrange for the tool to be detached from the boring rod at a
fixed point in each stroke, and this has led to the general

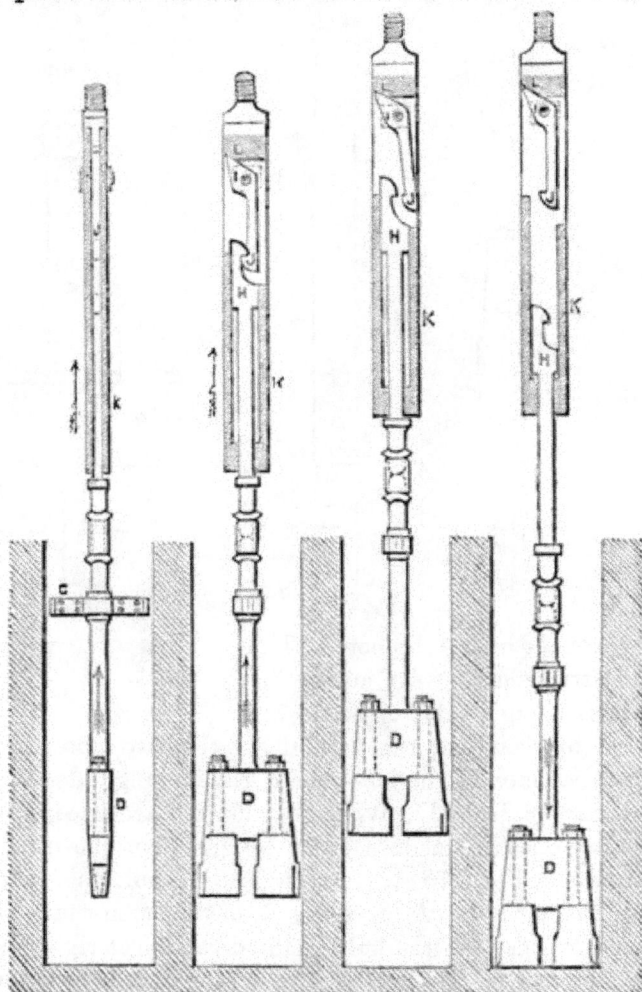

Fig. 178.   Fig. 179.   Fig. 180.   Fig. 181.

adoption of *free-falling tools*.   M. Dru's plan of self-acting free-
falling tool, liberated by reaction, is shown in side and front
view in Figs. 178 to 181.   The hook H, attached to the head

of the boring tool D, slides vertically in the box K, which is screwed to the lower extremity of the boring rod; and the hook engages with the catch J, centred in the sides of the box K, whereby the tool is lifted as the boring rod rises. The tail of the catch J bears against an inclined plane L, at the top of the box K; and the two holes carrying the centre-pin I of the catch, are made oval in the vertical direction, so as to allow a slight vertical movement of the catch. When the boring rod reaches the top of the stroke, it is stopped suddenly by the tail end of the beam B, Fig. 167, striking upon the wood buffer-block E; and the shock thus occasioned causes a slight jump of the catch J in the box K; the tail of the catch is thereby thrown outwards by the incline L, as shown in Fig. 180, liberating the hook H, and the tool then falls freely to the bottom of the bore-hole, as shown in Fig. 181. When the boring rod descends again after the tool, the catch J again engages with the hook H, enabling the tool to be raised for the next blow, as in Fig. 179.

Another construction of self-acting free-falling tool, liberated by a separate disengaging rod, is shown in side and front view in Figs. 182 to 186. This tool consists of four principal pieces, the hook H, the catch J, the pawl I, and the disengaging rod M. The hook H, carrying the boring tool D, slides between the two vertical sides of the box K, which is screwed to the bottom of the boring rod; and the catch J works in the same space upon a centre-pin fixed in the box, so that the tool is carried by the rod, when hooked on the catch, as shown in Fig. 183. At the same time the pawl I, at the back of the catch J, secures it from getting unhooked from the tool; but this pawl is centred in a separate sliding hoop N, forming the top of the disengaging rod M, which slides freely up and down within a fixed distance upon the box K; and in its lowest position the hoop N rests upon the upper of the two guides P P, Fig. 182, through which the disengaging rod M slides outside the box K. In lowering the boring rod, the disengaging rod M reaches the bottom of the bore-hole first, as shown in Figs. 182, 183, and being then stopped it prevents the pawl I from descending

any lower; and the inclined back of the catch J sliding down
past the pawl, the latter forces the catch out of the hook H, as

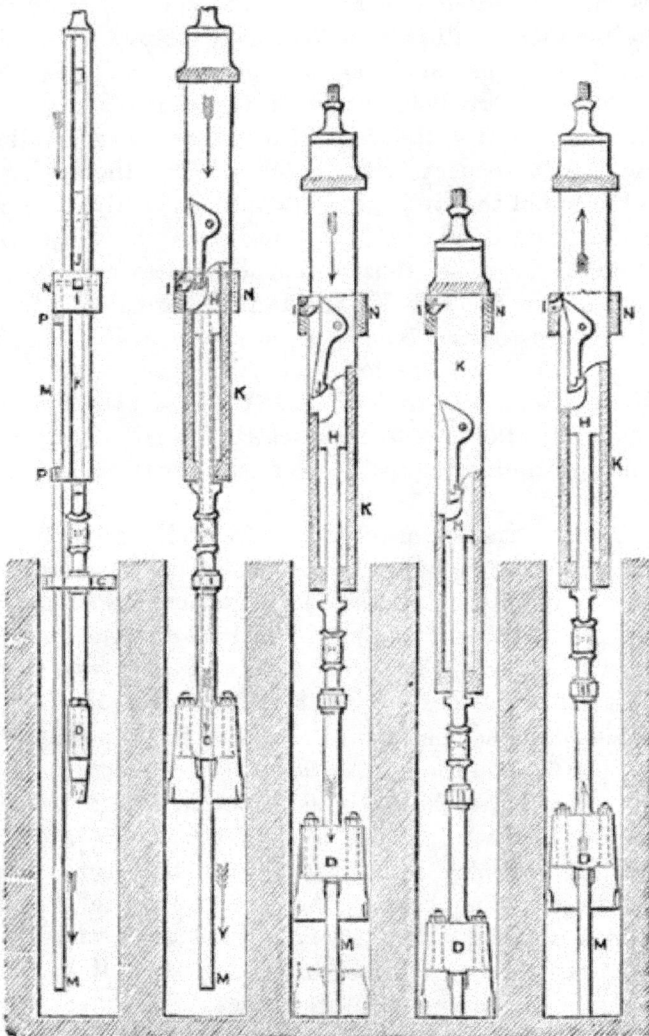

Fig. 182.   Fig. 183.   Fig. 184.   Fig. 185.   Fig. 186.

shown in Fig. 184, thus allowing the tool D to fall freely and
strike its blow. The height of fall of the tool is always the

same, being determined only by the length of the disengaging rod M.

The blow having been struck, and the boring rod continuing to be lowered to the bottom of the hole, the catch J falls back into its original position, and engages again with the hook H, as shown in Fig. 185, ready for lifting the tool in the next stroke. As the boring rod rises, the tail of the catch J trips up the pawl I in passing, as shown in Fig. 184, allowing the catch to pass freely; and the pawl before it begins to be lifted returns to the original position, shown in Fig. 185, where it locks the catch J, and prevents any risk of its becoming un-hooked either in raising or lowering the tool in the well.

The boring tool shown in Figs. 171, 172, which was employed for boring a well of 19 inches diameter, weighs $\frac{3}{4}$ ton, and is liberated by reaction, by the arrangement shown in Figs. 178 to 181; and the same mode of liberation was applied in the first instance to the larger tool, shown in Figs. 174 to 177, em-ployed in sinking a well of 47 in. diameter at Butte-aux-Cailles. The great weight of the latter tool, however, amounting to as much as $3\frac{1}{2}$ tons, necessitated so violent a shock, for the purpose of liberating the tool by reaction, that the boring rods and the rest of the apparatus would have been damaged by a continu-ance of that mode of working; and M. Dru was therefore led to design the arrangement of the disengaging rod for releasing the tool, as shown in Figs. 182, 183. In this case the cross-guide G fixed upon the tool is made with an eye for the disengaging rod M to work through freely. For borings of small diameter, however, the disengaging rod cannot supersede the reaction system of liberation, as the latter alone is able to work in borings as small as $3\frac{1}{4}$ inches diameter; and a bore-hole no larger than this diameter has been successfully completed by M. Dru with the reaction tool to a depth of 750 feet.

The boring rods employed are of two kinds, wrought iron and wood. The wood rods seen in Figs. 167, 187, are used for borings of large diameter, as they possess the advantage of having a larger section for stiffness without increasing the weight; and also when immersed in water the greater portion of their weight is floated. The wood for the rods requires to be carefully selected, and care has to be taken to choose the timber from the thick part of the tree, and not the toppings. In France, Lorraine or Vosges deals are preferred.

The boring rods, whether of wood or iron, are screwed together either by solid sockets, as in Fig. 189, or with separate collars, as in Figs. 188, 190. The separate collars are preferred for the purpose, on account of being easy to forge; and also because, as only one half of the collar works in coupling and uncoupling the rods, while the other half is fixed, the screw thread becomes worn only at one end, and by changing the collar, end for end, a new thread is obtained when one is worn out, the worn end being then jammed fast as the fixed end of the collar.

The boring rod is guided in the lower part of the hole by a lantern R, Fig. 167, shown to a larger scale in Fig. 187,

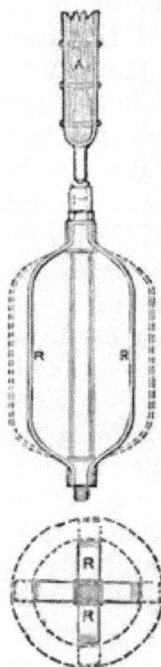

Fig. 187.

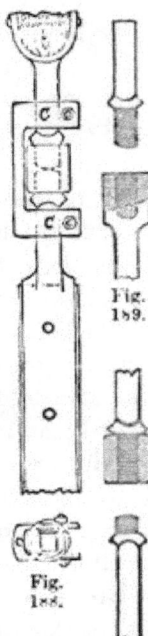

Fig. 189.

Fig. 188.

Fig. 190.

which consists of four vertical iron bars curved in at both ends, where they are secured by movable sockets upon the boring rod, and fixed by a nut at the top. By changing the bars, the size of the lantern is readily adjusted to any required diameter of bore-hole, as indicated by the dotted lines. In raising up or letting down the boring rod, two lengths of about 30 feet each are detached or added at once, and a few shorter rods of different lengths are used to make up the exact length required. The coupling screw S, Fig. 166, by which the boring rod is connected to the working beam B, serves to complete the adjustment of length; this is turned by a cross-bar, and then secured by a cross-pin through the screw.

In ordinary work, breakages of the boring rod generally take

place in the iron, and more particularly at the part screwed, as
that is the weakest part. In the case of breakages, the tools
usually employed for picking up the broken ends are a conical
screwed socket, shown in Fig. 191, and a crow's
foot, shown in Fig. 192 ; the socket being made
with an ordinary V-thread for cases where the
breakage occurs in the iron, but having a
sharper thread like a wood screw, when used
where the breakage is in one of the wood rods.
In order to ascertain the shape of the fractured
end left in the bore-hole, and its position relatively to the centre line of the hole, a similar conical
socket is first lowered, having its under surface filled
up level with wax, so as to take an impression of the
broken end, and show what size of screwed socket should be
employed for getting it up. Tools with nippers are sometimes
used in large borings, as it is not advisable to subject the rods
to a twist.

Fig. 191.

Fig. 192.

When the boring tool has detached a sufficient quantity of
material, the boring rod and tool are drawn up by means of the
rope O, Fig. 166, winding up the drum Q, which is driven by
straps and gearing from the steam-engine
T.   A shell is then lowered into the bore-
hole by the wire-rope U, from the other
drum V, and is afterwards drawn up again
with the excavated material. A friction
brake is applied to the drum Q, for regulating the rate of lowering the boring rod
down the well. The shell shown in Figs.
194, 195, consists of a riveted iron cylinder, with a handle at the top, which can
either be screwed to the boring rod or
attached to the wire rope ; and the bottom is closed by a large
valve, opening inwards. Two different forms of valve are used,
either a pair of flap-valves, as shown in Fig. 194, or a single-
cone valve Fig. 195 ; and the bottom ring of the cylinder,
forming the seating of the valve, is forged solid, and steeled on

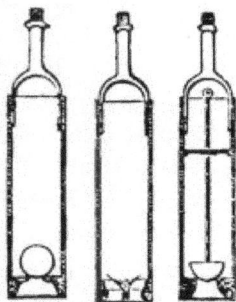

Fig. 193.   Fig. 194.   Fig. 195.

the lower edge. On lowering this cylinder to the bottom of
the bore-hole, the valve opens, and the loose material enters the
cylinder, where it is retained by the closing of the valve, whilst
the shell is drawn up again to the surface. In boring through
chalk, as in the case of the deep wells in the Paris basin, the
hole is first made of about half the final diameter for 60 to 90
feet depth, and it is then enlarged to the full diameter by using
a larger tool. This is done for convenience of working ; for if
the whole area were acted upon at once, it would involve crushing all the flints in the chalk ; but, by putting a shell in the
advanced hole, the flints that are detached during the working
of the second larger tool are received in the shell and removed
by it, without getting broken by the tool.

The resistance experienced in boring through different strata
is various; and some rocks passed through are so hard that
with 12,000 blows a day of a boring tool weighing nearly
10 cwt., with 19 inches height of fall, the bore-hole was advanced
only 3 to 4 inches a day. As the opposite case, strata of running sand have been met with so wet that a slight movement of
the rod at the bottom of the hole was sufficient to make the sand
rise 30 to 40 feet in the bore-hole. In these cases Dru has
adopted the Chinese method of effecting a speedy clearance, by
means of a shell closed by a large ball-clack at the bottom, as
shown in Fig. 193, and suspended by a rope, to which a vertical
movement is given ; each time the shell falls upon the sand a
portion of this is forced up into the cylinder, and retained there
by the ball-valve.

Borings of large diameter, for mines or other shafts, are also
sunk by means of the same description of boring tools, only
considerably increased in size, extending up to as much as
14 feet diameter. The well is then lined with cast-iron or
wrought-iron tubing, for the purpose of making it water-tight ;
and a special contrivance, invented by Kind, and alluded to at
p. 127, has been adopted for making a water-tight joint between
the tubing and the bottom of the well, or with another portion
of tubing previously lowered down. This is done by a stuffing
box, shown in Fig. 196, which contains a packing of moss at

A A. The upper portion of the tubing is drawn down to the lower portion by the tightening screws B B, so as to compress the moss-packing when the weight is not sufficient for the purpose. A space C is left be-
tween the tubing and the side
of the well, to admit of the
passage of the stuffing-box
flange, and also for running
in concrete for the completion
of the operation. The moss-
packing rests upon the bottom
flange D; but this flange is
sometimes omitted. The joint
is thus simply made by pressing out the moss-packing
against the sides of the well;
and this material, being easily
compressible and not likely to decay under water, is found to
make a very satisfactory and durable joint.

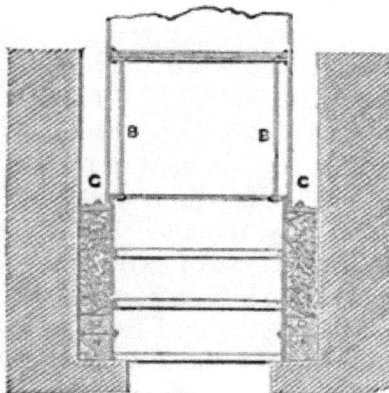

Fig. 196.

M. Dru states that the reaction tool has been successfully employed for borings up to as large as about 4 feet diameter, witness
the case of the well at Butte-aux-Cailles of 47 inches diameter;
but beyond that size he considers the shock requisite to liberate
the larger and heavier tool would probably be so excessive as
to be injurious to the boring rods and the rest of the attachments; and he therefore designed the arrangement of the disengaging rod for liberating the tool in borings of large
diameter, whereby all shock upon the boring rods was avoided
and the tool was liberated with complete certainty.

In practice it is necessary, as with the common chisel, to
turn the boring tool partly round between each stroke, so as to
prevent it from falling every time into the same position at the
bottom of the well; and this was effected in the well at Butte-
aux-Cailles by manual power at the top of the well, by means
of a long hand-lever fixed to the boring rod by a clip bolted on,
which was turned round by a couple of men through part of a
revolution during the time that the tool was being lifted. The

turning was ordinarily done in the right-hand direction only, so as to avoid the risk of unscrewing any of the screwed couplings of the boring rods; and care was taken to give the boring rod half a turn when the tool was at the bottom, so as to tighten the screw-couplings, which otherwise might shake loose. In the event of a fracture, however, leaving a considerable length of boring rod in the hole, it was sometimes necessary to have the means of unscrewing the couplings of the portion left in the hole, so as to raise it in parts instead of all at once. In that case a locking clip was added at each screwed joint above, and secured by bolts, as shown at C in Fig. 188, at the time of putting the rods together for lowering them down the well to recover the broken portion; and by this means the ends of the rods were prevented from becoming unscrewed in the coupling sockets, when the rods were turned round backwards for un-screwing the joints in the broken length at the bottom of the bore-hole.

When running sands are met with, the plan adopted is to use the Chinese ball-scoop, or shell, Fig. 193, described for clear-ing the bottom of the bore-hole; and where there is too much sand for it to be got rid of in this way, a tube has to be sent down from the surface to shut off the sand. This, of course, necessitates diminishing the diameter of the hole in passing through the sand; but on reaching the solid rock below the running sand, an expanding tool is used for continuing the bore-hole below the tubing with the same diameter as above it, so as to allow the tubing to go down with the hole.

In the case of meeting with a surface of very hard rock at a considerable inclination to the bore-hole, M. Dru employs a tool, the cutters of which are fixed in a circle all round the edge of the tool, instead of in a single diameter line; the length of the tool is also considerably increased in such cases, as compared with the tools used for ordinary work, so that it is guided for a length of as much as 20 feet. He uses this tool in all cases where from any cause the hole is found to be going crooked, and has even succeeded by this means in straightening a hole that had previously been bored crooked.

The cutting action of this tool is all round its edge; and therefore in meeting with an inclined hard surface, as there is nothing to cut on the lower side, the force of the blow is brought to bear on the upper side alone, until an entrance is effected into the hard rock in a true straight line with the upper part of the hole.

Although as regards diameter, depth, and flow of water in favourable localities, some extraordinary results have been obtained with this system of boring by rods worked by steam power, yet, as Dru himself observes, " in some instances his own experience of boring had been, that owing to the difficulties attending the operation, the occurrence of delays from accidents was the rule, while the regular working of the machinery was the exception." A further disadvantage to be noticed is that, owing to the time and labour involved in raising and lowering heavy rods in borings of 10 inches diameter and upwards, there is a strong inducement to keep the boring tool at work, for a much longer period than is actually necessary for breaking up fresh material at each stroke. The fact is that after from 100 to 200 blows have been given, the boring tool merely falls into the accumulated *débris* and pounds this into dust, without again touching the surface of the solid rock. It may therefore be easily understood how much time is totally lost out of the periods of five to eight hours during which, with the rod system, the tool is allowed to continue working.

MATHER AND PLATT'S SYSTEM.

In Mather and Platt's method of boring adopted in England, the rope has been reverted to, in place of the iron or wood rods used on the Continent. A flexible rope admits of being handled with greater facility than iron rods, but wants the advantage of rigidity : in the Chinese method it admitted of withdrawing the chisel or bucket very rapidly, but gave no certainty to the operation of the chisel at the bottom of the hole. The rods on the other hand enable a very effective blow to be given, with a definite turning or screwing motion between the blows according to the requirements of the strata; but the time and trouble of raising heavy rods from great depths on each occasion of changing from boring to clearing out the hole form a serious drawback, which makes the stoppages occupy really a longer time than the actual working of the machinery.

The method invented by Colin Mather, and manufactured by Mather and Platt, of Oldham, employed largely in England for deep boring, seems to combine the advantages of the systems hitherto used, and to be free from many of their disadvantages. The distinctive fea-

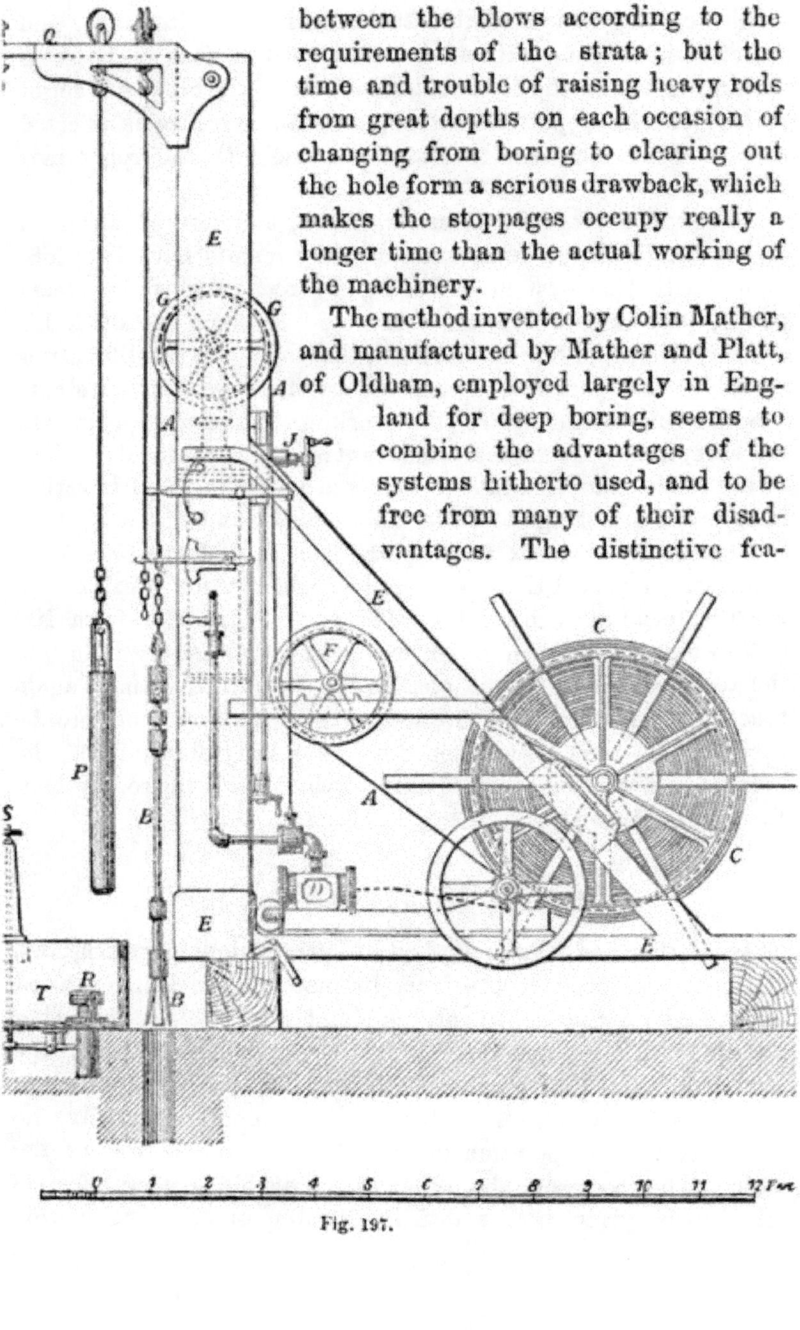

Fig. 197.

tures of this plan, which is shown in Figs. 197 to 204, are the
mode of giving the percussive action to the boring tool, and the

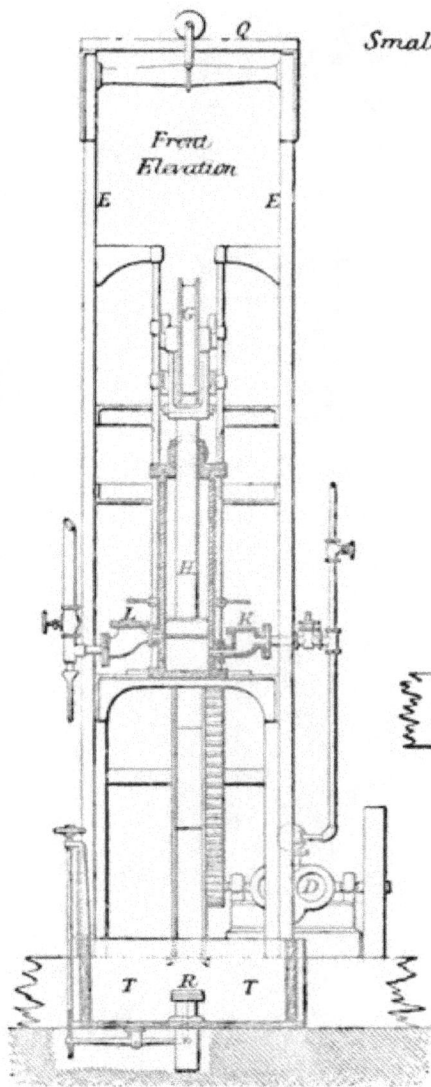

Small Boring Machine

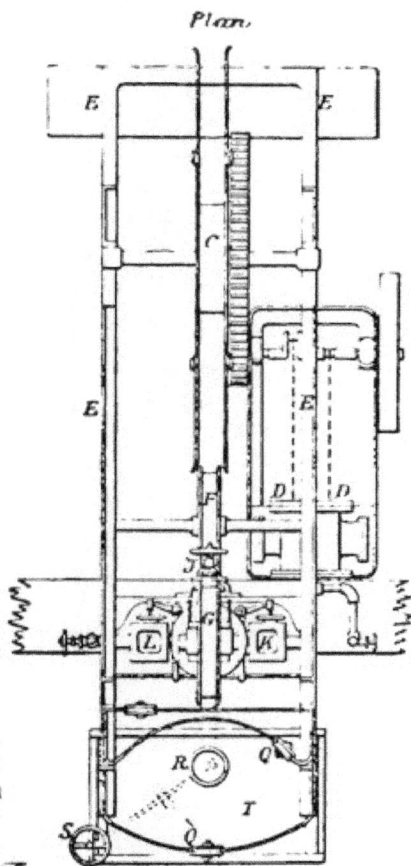

Fig. 198.   Fig. 199.

construction of the tool or boring-head, and of the shell-pump

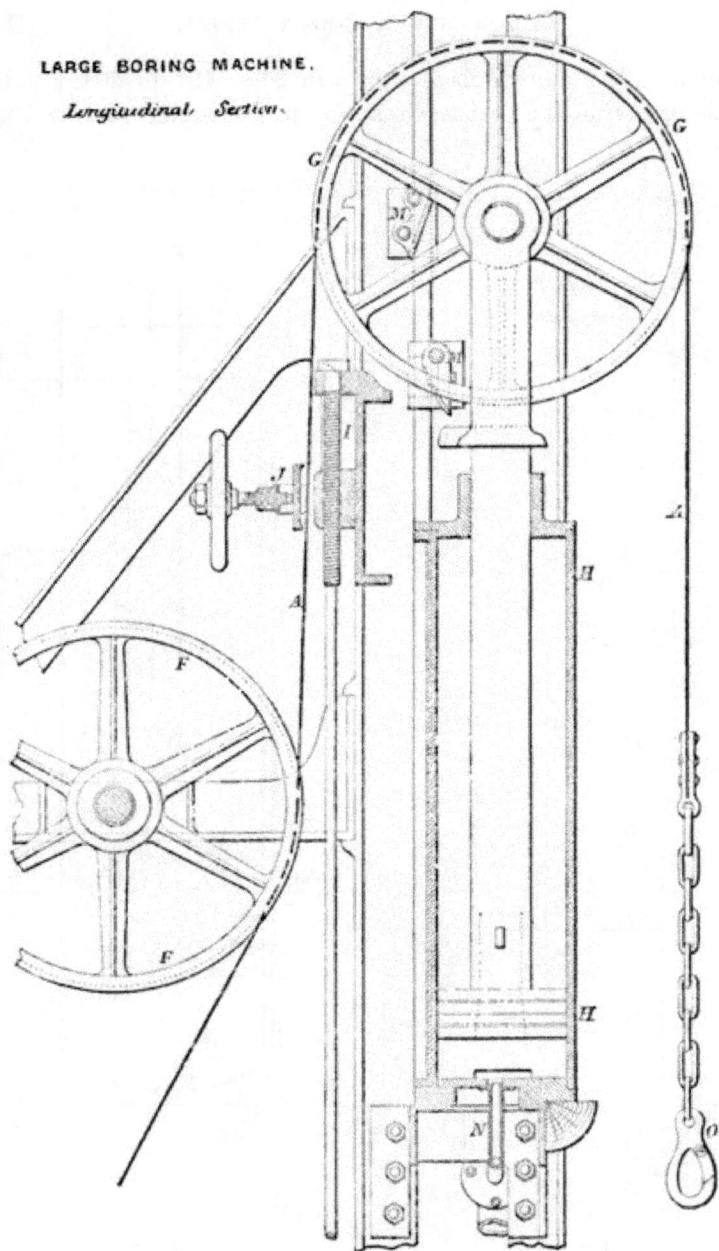

LARGE BORING MACHINE.

Longitudinal Section.

Fig 200.

for clearing out the hole after the action of the boring head. Instead of these implements being attached to rods, they are suspended by a flat hemp rope, about $\frac{1}{2}$ inch thick and $4\frac{1}{2}$ inches broad, such as is commonly used at collieries ; and the boring tool and shell-pump are raised and lowered as quickly in the bore-hole as the bucket and cages in a colliery shaft.

The flat rope A A, Fig. 197, from which the boring head B is suspended, is wound upon a large drum C driven by a steam-engine D with a reversing motion, so that one man can regulate the operation with the greatest ease. All the working parts are fitted into a wood or iron framing E E, rendering the whole a compact and complete machine. On leaving the drum C the rope passes under a guide pulley F, and then over a large pulley G carried in a fork at the top of the piston rod of a vertical single-acting steam cylinder.

This cylinder, by which the percussive action of the boring head is produced, is shown to a larger scale in the vertical sections, Figs. 200, 201 ; and in the larger size of machine here shown, the cylinder is fitted with a piston of 15 inches diameter, having a heavy cast-iron rod 7 inches square, which is made with a fork at the top carrying the flanged pulley G of about 3 feet diameter and of sufficient breadth for the flat rope A to pass over it. The boring-head having been lowered by the winding drum to the bottom of the bore-hole, the rope is fixed secure at that length by the clamp J ; steam is then admitted underneath the piston in the cylinder H by the steam valve K, and the boring tool is lifted by the ascent of the piston rod and pulley G ; and on arriving at the top of the stroke the exhaust valve L is opened for the steam to escape, allowing the piston rod and carrying pulley to fall freely with the boring tool, which falls with its full weight to the bottom of the bore-hole. The exhaust port is 6 inches above the bottom of the cylinder, while the steam port is situated at the bottom; and there is thus always an elastic cushion of steam retained in the cylinder of that thickness for the piston to fall upon, preventing the piston from striking the bottom of the cylinder. The steam and exhaust valves are worked with a self-acting motion by the

Large Boring Machine.

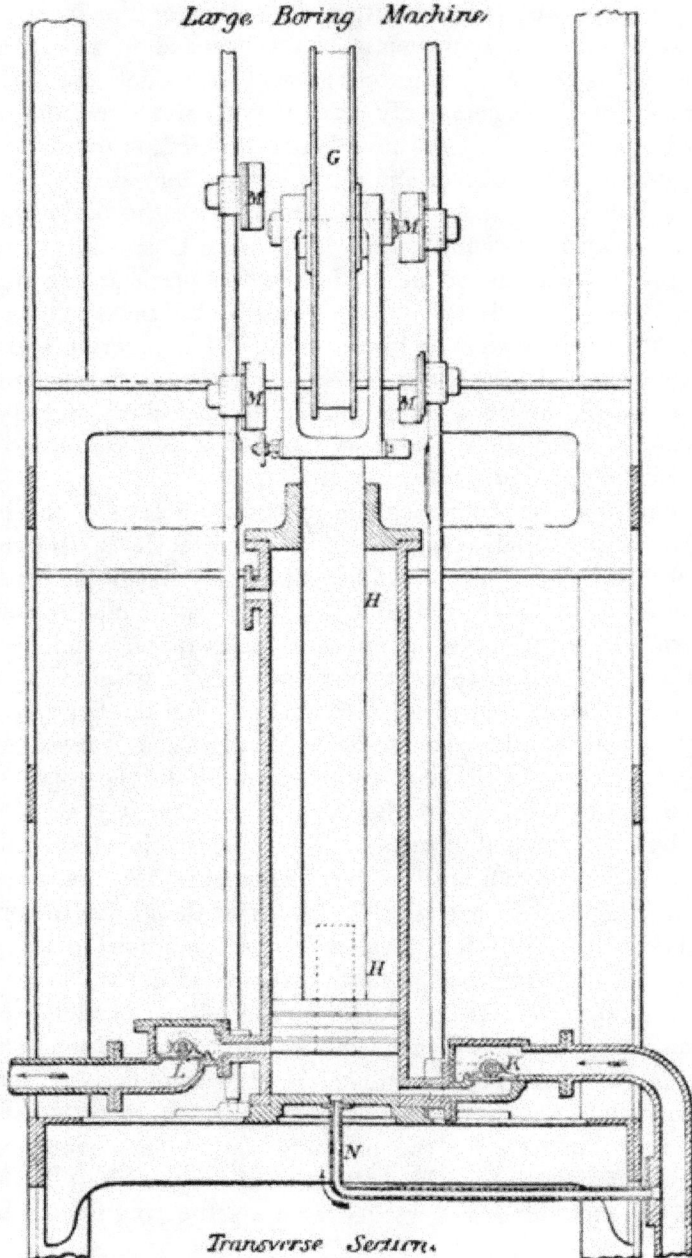

Transverse Section.

Fig. 201.

tappets M M, which are actuated by the movement of the piston rod; and a rapid succession of blows is thus given by the boring tool on the bottom of the bore-hole. As it is necessary that motion should be given to the piston before the valves can be acted upon, a small jet of steam N is allowed to be constantly blowing into the bottom of the cylinder; this causes the piston to move slowly at first, so as to take up the slack of the rope and allow it to receive the weight of the boring head gradually and without a jerk. An arm attached to the piston-rod then comes in contact with a tappet which opens the steam valve K, and the piston rises quickly to the top of the stroke; another tappet worked by the same arm then shuts off the steam, and the exhaust valve L is opened by a corresponding arrangement on the opposite side of the piston rod, as shown in Fig. 201. By shifting these tappets the length of stroke of the piston can be varied from 1 to 8 feet in the large machine, according to the material to be bored through; and the height of fall of the boring head at the bottom of the bore-hole is double the length of stroke of the piston. The fall of the boring head and piston can also be regulated by a weighted valve on the exhaust pipe, checking the escape of the steam, so as to cause the descent to take place slowly or quickly, as may be desired.

The boring head B, Fig. 197, is shown to a larger scale in Figs. 202, 203, and consists of a wrought-iron bar about 4 inches diameter and 8 feet long, to the bottom of which a cast-iron cylindrical block C is secured. This block has numerous square holes through it, into which the chisels or cutters D D are inserted with taper shanks, as shown in Fig. 203, so as to be very firm when working, but to be readily taken out for repairing and sharpening. Two different arrangements of the cutters are shown in the elevation, Fig. 202, and the plan, Fig. 204. A little above the block C another cylindrical casting E is fixed upon the bar B, which acts simply as a guide to keep the bar perpendicular. Higher still is fixed a second guide F, but on the circumference of this are secured cast-iron plates made with ribs of a saw-tooth or ratchet shape, catching only in one direction; these ribs are placed at an inclination like seg-

Fig. 203.

Boring Head.

Elevation Sectional Elevation

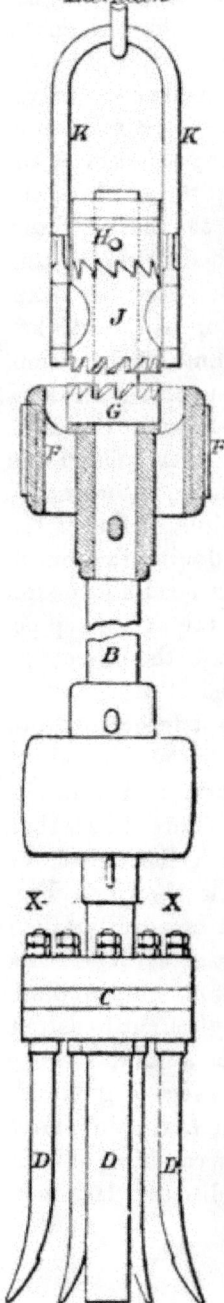

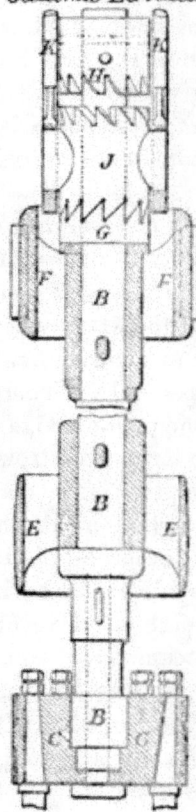

Plan of X.X.

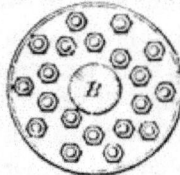

Plan at Bottom inverted.

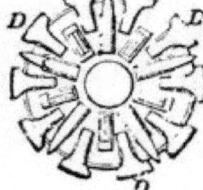

Fig. 202. Fig. 204.

ments of a screw-thread of very long pitch, so that as the guide bears against the rough sides of the bore-hole when the bar is raised or lowered they assist in turning it, for causing the cutters to strike in a fresh place at each stroke. Each alternate plate has the projecting ribs inclined in the opposite direction, so that one half of the ribs are acting to turn the bar round in rising, and the other half to turn it in the same direction in falling. These projecting spiral ribs simply assist in turning the bar, and immediately above the upper guide F is the arrangement by which the definite rotation is secured. To effect this object two cast-iron collars, G and H, are cottered fast to the top of the bar B, and placed about 12 inches apart; the upper face of the lower collar G is formed with deep ratchet-teeth of about 2 inches pitch, and the under face of the top collar H is formed with similar ratchet-teeth, set exactly in line with those on the lower collar. Between these collars and sliding freely on the neck of the boring bar B is a deep bush J, which is also formed with corresponding ratchet-teeth on both its upper and lower faces; but the teeth on the upper face are set half a tooth in advance of those on the lower face, so that the perpendicular side of each tooth on the upper face of the bush is directly above the centre of the inclined side of a tooth on the lower face. To this bush is attached the wrought-iron bow K, by which the whole boring bar is suspended with a hook and shackle O, Fig. 200, from the end of the flat rope A. The rotary motion of the bar is obtained as follows: when the boring tool falls and strikes the blow, the lifting bush J, which during the lifting has been engaged with the ratchet-teeth of the top collar H, falls upon those of the bottom collar G, and thereby receives a twist backwards through the space of half a tooth; and on commencing to lift again, the bush rising up against the ratchet-teeth of the top collar H, receives a further twist backwards through half a tooth. The flat rope is thus twisted backwards to the extent of one tooth of the ratchet; and during the lifting of the tool it untwists itself again, thereby rotating the boring tool forwards through that extent of twist between each successive blow of the tool. The amount of the rotation may be varied by making the ratchet-teeth of coarser or finer pitch. The motion is entirely self-acting, and the rotary movement of the boring tool is ensured with mechanical accuracy. This simple and most effective action taking place at every blow of the tool produces a constant change in the position of the cutters, thus increasing their effect in breaking the rock.

The shell-pump, for raising the material broken up by the boring head, is shown in Figs. 205, 206, and consists of a cylindrical shell or barrel P of cast iron, about eight feet long and a little smaller in diameter than the size of the bore-hole. At the bottom is a clack A opening upwards, somewhat similar to that in ordinary pumps; but its seating, instead of being fastened to the cylinder P, is in an annular frame C, which is held up against the bottom of the cylinder by a rod D passing up to

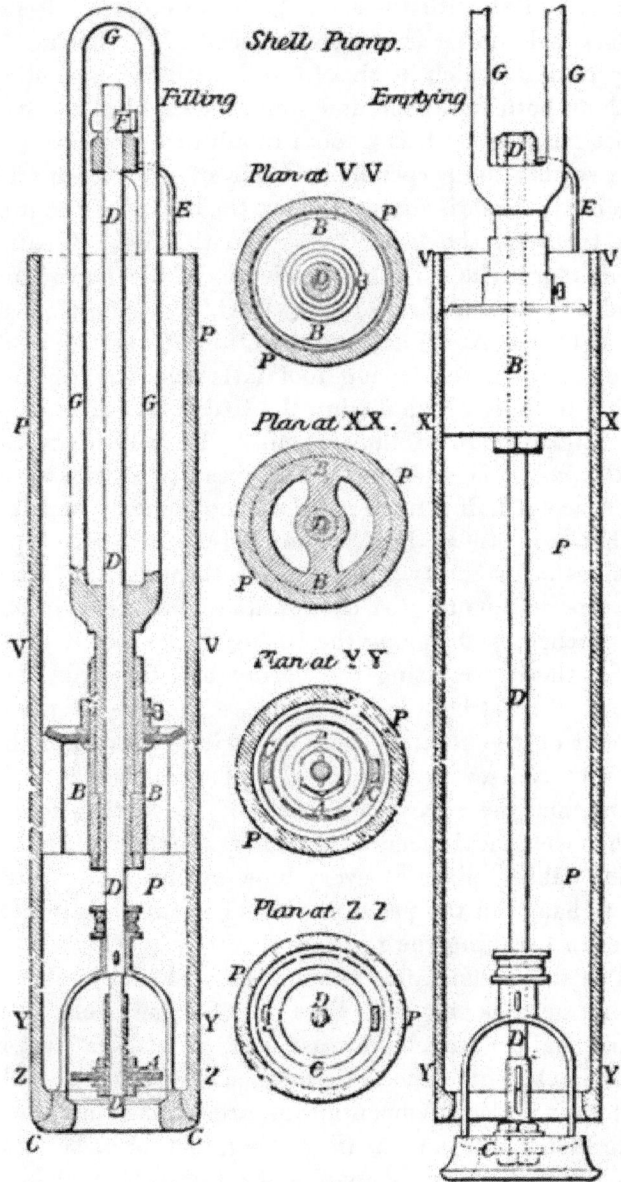

Shell Pump.

Filling

Emptying

Plan at V V

Plan at X X.

Plan at Y Y

Plan at Z Z

Fig. 205.

Fig. 206.

a wrought-iron guide E at the top, where it is secured by a
cotter F.   Inside the cylinder works a bucket B, similar to that
of a common lift-pump, having an indiarubber disc valve on
the top side ; and the rod D of the bottom clack passes freely
through the bucket.   The rod G of the bucket itself is formed
like a long link in a chain, and by this link the pump is suspended from the shackle O, Fig. 200, at the end of the flat rope,
the guide E, Fig. 205, preventing the bucket from being drawn
out of the cylinder.   The bottom clack A is made with an
indiarubber disc, which opens sufficiently to allow the water
and smaller particles of stone to enter the cylinder ; and in order
to enable the pieces of broken rock to be brought up as large as
possible, the entire clack is free to rise bodily about 6 inches
from the annular frame C, as shown in Fig. 205, thereby affording ample space for large pieces of rock to enter the cylinder,
when drawn in by the up stroke of the bucket.

The general working of the boring machine is as follows.
The winding drum C, Fig. 197, is 10 feet diameter in the large
machine, and is capable of holding 3000 feet length of rope
$4\frac{1}{2}$ inches broad and $\frac{1}{2}$ inch thick.   When the boring head B is
hooked on the shackle at the end of the rope A, its weight pulls
round the drum and winding engine, and by means of a break
it is lowered steadily to the bottom of the bore-hole ; the rope
is then secured at that length by screwing up tight the clamp J.
The small steam jet N, Figs. 200, 201, is next turned on, for
starting the working of the percussion cylinder H ; and the
boring head is then kept continually at work, until it has broken
up a sufficient quantity of material at the bottom of the bore-
hole.   The clamp J which grips the rope is made with a slide
and screw I, Fig. 200, whereby more rope can be gradually given
out as the boring head penetrates deeper in the hole.   In order
to increase the lift of the boring head, or to compensate for the
elastic stretching of the rope, which is found to amount to
1 inch in each 100 feet length, it is simply necessary to raise
the top pair of tappets on the tappet rods whilst the percussive motion is in operation.   When the boring head has
been kept at work long enough, the steam is shut off from the

percussion cylinder, the rope unclamped, the winding engine put in motion, and the boring head wound up to the surface, where it is then slung from an overhead suspension bar Q, Fig. 197, by means of a hook mounted on a roller for running the boring head away to one side, clear of the bore-hole.

The shell-pump is next lowered down the bore-hole, by the rope, and the *débris* pumped into it by lowering and raising the bucket about three times at the bottom of the hole, which is readily effected by means of the reversing motion of the winding engine. The pump is then brought up to the surface, and emptied by the following very simple arrangement : it is slung by a traversing hook from the overhead suspension bar Q, Fig. 197, and is brought perpendicularly over a small table R in the waste tank T ; and the table is raised by the screw S until it receives the weight of the pump. The cotter F, Fig. 205, which holds up the clack seating C at the bottom of the pump, is then knocked out ; and the table being lowered by the screw, the whole clack seating C descends with it, as shown in Fig. 206, and the contents of the pump are washed out by the rush of water contained in the pump cylinder. The table is then raised again by the screw, replacing the clack seating in its proper position, in which it is secured by driving the cotter F into the slot at the top ; and the pump is again ready to be lowered down the bore-hole as before. It is sometimes necessary for the pump to be emptied and lowered three or four times in order to remove all the material that has been broken up by the boring-head at one operation.

The rapidity with which these operations may be carried on is found in the experience of the working of the machine to be as follows. The boring head is lowered at the rate of 500 feet a minute. The percussive motion gives twenty-four blows a minute ; this rate of working continued for about ten minutes in red sandstone and similar strata is sufficient for enabling the cutters to penetrate about 6 inches depth, when the boring head is wound up again at the rate of 300 feet a minute. The shell-pump is lowered and raised at the same speeds, but only remains down about two minutes ; and the emptying of the pump when drawn up occupies about two or three minutes.

In the construction of this machine it will be seen that the great desideratum of all earth boring has been well kept in view; namely, to bore holes of large diameter to great depths with rapidity and safety. The object is to keep either the boring head or the shell-pump constantly at work at the bottom of the bore-hole, where the actual work has to be done; to lose as little time as possible in raising, lowering, and changing the tools; to expedite all the operations at the surface; and to economise manual labour in every particular. With this machine, one man standing on a platform at the side of the percussion cylinder performs all the operations of raising and lowering by the winding engine, changing the boring-head and shell pump, regulating the percussive action, and clamping or unclamping the rope: all the handles for the various steam valves are close to his hand, and the brake for lowering is worked by his foot. Two labourers attend to changing the cutters and clearing the pump. Duplicate boring heads and pumps are slung to the overhead suspension bar Q, Fig. 197, ready for use, thus avoiding all delay when any change is requisite.

As is well known by those who have charge of such operations, in well boring innumerable accidents and stoppages occur from causes which cannot be prevented, with however much vigilance and skill the operations may be conducted. Hard and soft strata intermingled, highly inclined rocks, running sands, and fissures and dislocations are fruitful sources of annoyance and delay, and sometimes of complete failure; and it will therefore be interesting to notice a few of the ordinary difficulties arising out of these circumstances. In all the bore-holes yet executed by this system, the various special instruments used under any circumstances of accident or complicated strata are fully shown in Figs. 207 to 215.

The boring head while at work may suddenly be jammed fast, either by breaking into a fissure, or in consequence of broken rock falling upon it from loose strata above. All the strain possible is then put upon the rope, either by the percussion cylinder or by the winding engine; and if the rope is an old one or rotten it breaks, leaving perhaps a long length in the

Fig. 208.

ACCIDENT TOOLS

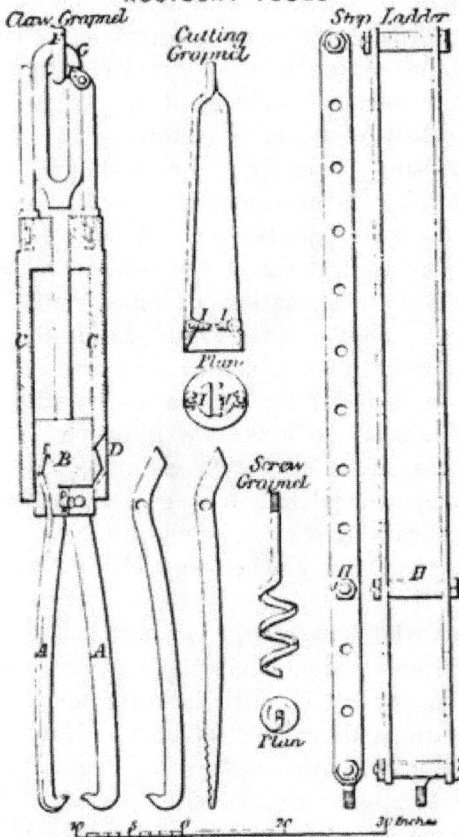

Claw Grapnel

Cutting Grapnel

Step Ladder

Breaking-up Bar.

Grapnel for Tubes.

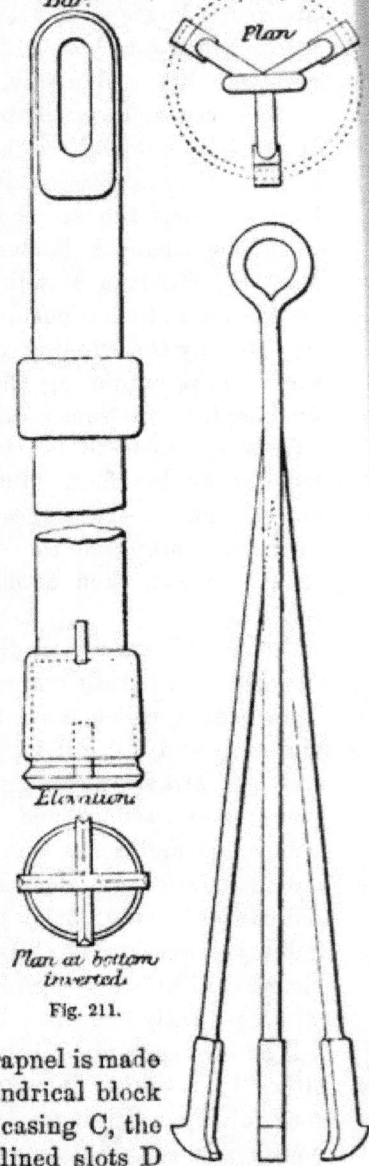

Plan

Screw Grapnel

Plan

Plan

Plan

Elevation

Plan at bottom inverted.

Fig. 211.

Fig. 207.   Fig. 210.

hole. The claw grapnel, shown in
Fig. 207, is then attached to the rope
remaining on the winding drum, and
is lowered until it rests upon the slack
broken rope in the bore-hole. The grapnel is made
with three claws A A centred in a cylindrical block
B, which slides vertically within the casing C, the
tail ends of the claws fitting into inclined slots D
in the casing. During the lowering of the grapnel,

Elevation

Fig. 212.

Grapnel for Cores

Plan at top

Straightening
Plug
for Tubes

Plan

GRAPNEL
FOR
STIFF CLAY.

Plan at Bottom

Section
of Bottom

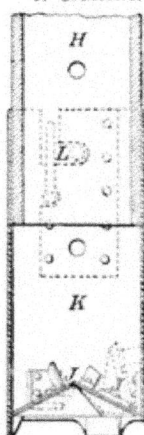

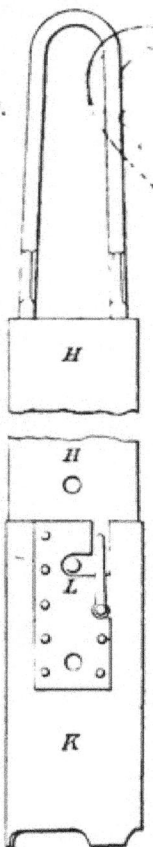

Elevation

Fig. 213.

Section of bottom

Fig 214.

Fig. 215.

the claws are kept open, in consequence of the trigger E being held up in the position shown in Fig. 207, by the long link F, which suspends the grapnel from the top rope. But as soon as the grapnel rests upon the broken rope below, the suspending link F continuing to descend allows the trigger E to fall out of it, and then in hauling up again, the

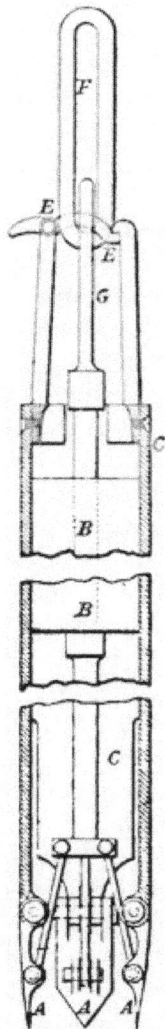

grapnel is lifted only by the bow G of the internal block B, and
the entire weight of the external casing C bears upon the inclined tail ends of the claws A, causing them to close in tight
upon the broken rope and lay hold of it securely. The claws
are made either hooked at the extremity or serrated. The grapnel is then hauled up sufficiently to pull the broken rope tight,
and wrought-iron rods 1 inch square with hooks attached at the
bottom are let down to catch the bow of the boring head, which
is readily accomplished. Two powerful screw-jacks are applied
to the rods at the surface, by means of the step-ladder shown in
Fig 209, in which the cross-pin H is inserted at any pair of the
holes, so as to suit the height of the screw-jacks.

If the boring head does not yield quickly to these efforts, the
attempt to recover it is abandoned, and it is got out of the way
by being broken up into pieces. For this purpose the broken
rope in the bore-hole has first to be removed, and it is therefore
caught hold of with a sharp hook and pulled tight in the hole,
while the cutting grapnel, shown in Fig. 208, is slipped over it
and lowered by the rods to the bottom. This tool is made with
a pair of sharp cutting jaws or knives I I opening upwards,
which in lowering pass down freely over the rope ; but when the
rods are pulled up with considerable force, the jaws nipping the
rope between them cut it through, and it is thus removed altogether from the bore-hole. The solid wrought-iron breaking-up
bar, Fig. 203, which weighs about a ton, is then lowered, and
by means of the percussion cylinder it is made to pound away
at the boring head, until the latter is either driven out of the
way into one side of the bore-hole, or broken up into such fragments as that, partly by the shell-pump and partly by the
grapnels, the whole obstacle is removed. The boring is then
proceeded with again, the same as before the accident.

The same mishap may occur with the shell-pump getting
jammed fast into the bore-hole, as illustrated in Fig. 216 ; and the
same means of removing the obstacle are then adopted. Experience has shown the danger of putting any greater strain
upon the rope than the percussion cylinder can exert ; and it is
therefore usual to lower the grapnel rods at once, if the boring

head or pump gets fast, thus avoiding the risk of breaking the rope.

The breaking of a cutter in the boring head is not an uncommon occurrence. If, however, the bucket grapnel, or the small screw grapnel, Fig. 210, be employed for its recovery, the hole is usually cleared without any important delay. The screw grapnel, Fig. 210, is applied by means of the iron grappling rods, so that by turning the rods the screw works itself round the cutter or other similar article in the bore-hole, and securely holds it while the rods are drawn up again to the surface. The bucket grapnel, Fig. 214, is also employed for raising clay, as well as for the purpose of bringing up cores out of the bore-hole, where these are not raised by the boring head itself in the manner already described. The action of this grapnel is nearly similar to that of the claw grapnel, Fig. 207 ; the three jaws A A, hinged to the bottom of the cylindrical casing C, and attached by connecting rods to the internal block B sliding within the casing C, are kept open during the lowering of the tool, the trigger E

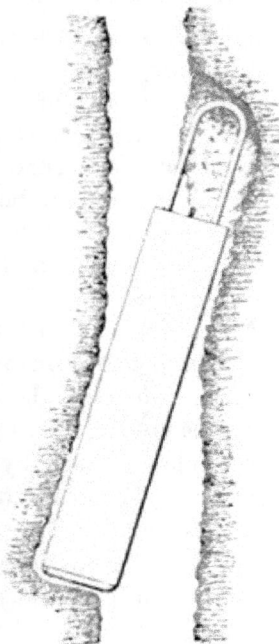

Shell Pump Jammed in Borehole

Fig. 216.

being held up in the position shown in Fig. 214, by the long suspending link F. On reaching the bottom, the trigger is liberated by the further descent of the link F, which, in hauling up again, lifts only the bow G of the internal block B; so that the jaws A are made to close inwards upon the core, which is thus grasped firmly between them and brought up within the grapnel. Where there is clay or similar material at the bottom of the bore-hole, the weight of the heavy block B in the grapnel causes the sharp edges of the pointed jaws to penetrate to some

depth into the material, a quantity of which is thus enclosed
within them and brought up.

Another grapnel which is also used where a bore-hole passes
through a bed of very stiff clay is shown in Fig. 215, and
consists of a long cast-iron cylinder H fitted with a sheet-iron
mouthpiece K at the bottom, in which are hinged three conical
steel jaws J J opening upwards. The weight of the tool forces
it down into the clay with the jaws open ; and then on raising
it the jaws, having a tendency to fall, cut into the clay and
enclose a quantity of it inside the mouthpiece, which on being
brought up to the surface is detached from the cylinder H and
cleaned out. A second mouthpiece is put on and sent down for
working in the bore-hole while the first is being emptied, the
attachment of the mouthpiece to the cylinder being made by a
common bayonet-joint D, so as to admit of readily connecting
and disconnecting it.

A running sand in soft clay is, however, the most serious
difficulty met with in well boring. Under such circumstances
the bore-hole has to be tubed from top to bottom, which greatly
increases the expense of the undertaking, not only by the cost
of the tubes, but also by the time and labour expended in inserting them. When a permanent water supply is the main object
of the boring, the additional expense of tubing the bore-hole is
not of much consequence, as the tubed hole is more durable,
and the surface water is thereby excluded ; but in exploring for
mineral it is a serious matter, as the final result of the bore-hole
is then by no means certain. The mode of inserting tubes has
become a question of great importance in connection with this
system of boring, and much time and thought having been spent
in perfecting the method now adopted, its value has been proved
by the repeated success with which it has been carried out.

The tubes used by Mather and Platt are of cast iron, varying
in thickness from $\frac{5}{8}$ to 1 inch according to their diameter, and
are all 9 feet in length. The successive lengths are connected
together by means of wrought-iron covering hoops 9 inches
long, made of the same outside diameter as the tube, so as to be
flush with it. These hoops are from $\frac{1}{4}$ to $\frac{3}{8}$ inch thick, and the

ends of each tube are reduced in diameter by turning down for 4½ inches from the end, to fit inside the hoops, as shown in Fig. 217. A hoop is shrunk fast on one end of each tube, leaving 4½ inches of socket projecting to receive the end of the next tube to be connected. Four or six rows of screws with countersunk heads, placed at equal distances round the hoop, are screwed through into the tubes to couple the two lengths securely together. Thus a flush joint is obtained both inside and outside the tubes. The lowest tube is provided at the bottom with a steel shoe, having a sharp edge for penetrating the ground more readily.

Tubing for Borehole

Section of Shoe of Tubing.

Fig. 217.

In small borings from 6 to 12 inches diameter, the tubes are inserted into the bore-hole by means of screw-jacks, by the simple and inexpensive method shown in Figs. 218, 219. The boring machine foundation A A, which is of timber, is weighted at B B by stones, pig iron, or any available material ; and two screw-jacks C C, each of about 10 tons power, are secured with the screws downwards, underneath the beams D D crossing the shallow well E, which is always excavated at the top of the bore-hole. A tube F having been lowered into the mouth of the bore-hole by the winding engine, a pair of deep clamps G are screwed tightly round it, and the screw-jacks acting upon these clamps force the tube down into the ground. The boring is then resumed, and as it proceeds the jacks are occasionally worked, so as to force the tube if possible even ahead of the boring tool. The clamps are then slackened and shifted up the tubes, to suit the length of the screws of the jacks ; two men work the jacks, and couple the lengths of tubes as they are successively added. The actual boring is carried on simultaneously within the tubes, and is not in the least impeded

M

by their insertion, which simply involves the labour of an
additional man or two.

Fig. 218.

Tube-Forcing Apparatus with Screwjacks.
Side Elevation.

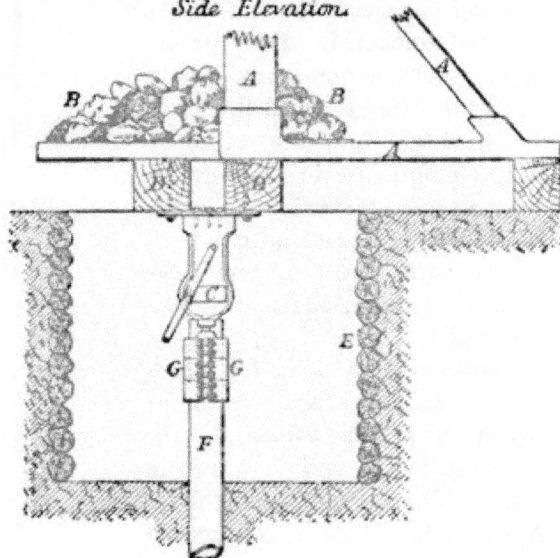

Front Elevation.

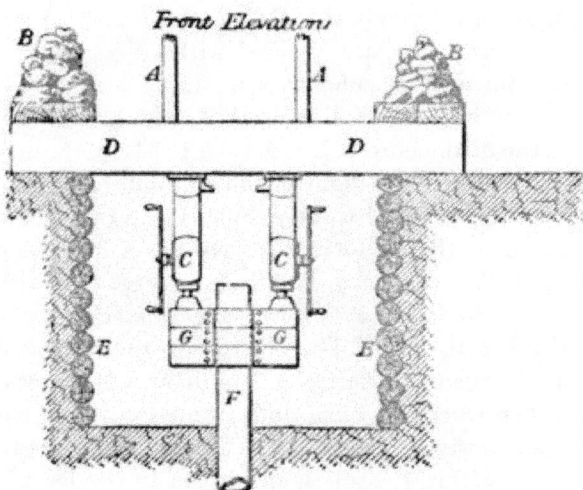

Fig. 219.

A more perfect and powerful tube-forcing apparatus is adopted where tubes of from 18 to 24 inches diameter have to be inserted to a great depth, an illustration of which is afforded by an extensive piece of work at the Horse Fort, standing in the channel at Gosport. This fort is a huge round tower, as shown in Fig. 220 ; and to supply the garrison with fresh water, a bore-hole is sunk into the chalk. A cast-iron well A, consisting of cylinders 6 feet diameter, and 5 feet long, has been sunk 90 feet into the bed of the channel in the centre of the fort, and from the bottom of this well is an 18-inch bore-hole B, which is tubed the whole distance with cast-iron tubes 1 inch thick, coupled as before described.

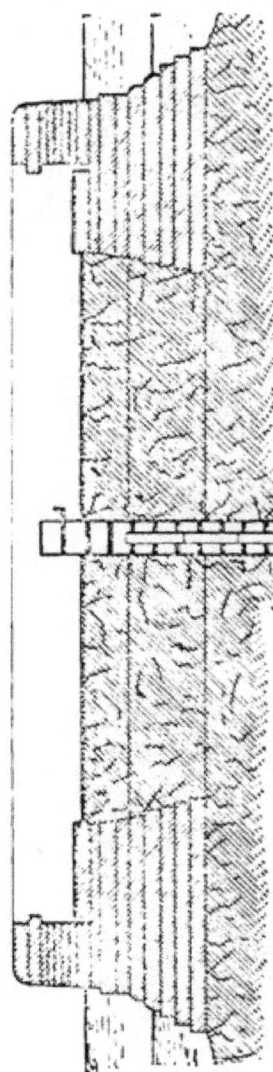

The method of inserting these tubes is shown in Fig. 222. Two wrought-iron columns C C, 6 inches diameter, are firmly secured in the position shown, by castings bolted to the flanges of the cylinders A A forming the well, so that the two columns are perfectly rigid and parallel to each other. A casting D, carrying on its under side two 5-inch hydraulic rams, I I of 4 feet length, is formed so as to slide freely between the columns,

M 2

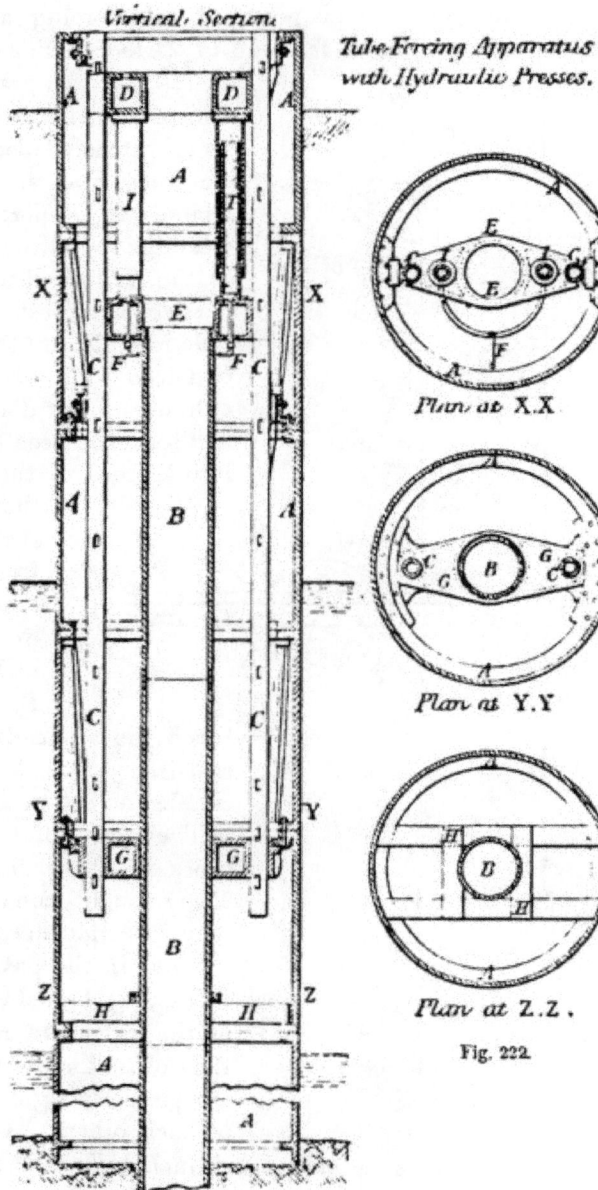

Vertical Section.

Tube-Forcing Apparatus
with Hydraulic Presses.

Plan at X.X.

Plan at Y.Y

Plan at Z.Z.

Fig. 222.

Fig. 221.

which act as guides; the hole in the centre of this casting is large enough to pass a bore-tube freely through it, and by means of cotters passed through the slots in the columns the casting is securely fixed at any height. A second casting E, exactly the same shape as the top one, is placed upon the top of the tubes B B to be forced down, a loose wrought-iron hoop being first put upon the shoulder at the top of the tube, large enough to prevent the casting E from sliding down the outside of the tubes; this casting or crosshead rests unsecured on the top of the tube and is free to move with it. The hydraulic cylinders I, with their rams pushed home, are lowered upon the crosshead E, and the top casting D to which they are attached is then secured firmly to the columns C by cottering through the slots. A small pipe F, having a long telescope joint, connects the hydraulic cylinders I with the pumps at the surface which supply the hydraulic pressure.

By this arrangement a force of 3 tons on the square inch, or about 120 tons total upon the two rams, has frequently been exerted to force down the tubes at the Horse Fort. After the rams have made their full stroke of about 3 feet 6 inches, the pressure is let off, and the hydraulic cylinders I with the top casting D, slide down the rams resting on the crosshead E, until the rams are again pushed home. The top casting D is then fixed in its new position upon the columns C, by cottering fast as before, and the hydraulic pressure is again applied; and this is repeated until the length of two tubes, making 18 feet, has been forced down. The whole hydraulic apparatus is then drawn up again to the top, another 18 feet of tubing added, and the operation of forcing down resumed. The tubes are steadied by guides at G and H, Fig. 221, shown also in the plans, Fig. 222.

The boring operations are carried on uninterruptedly during the process of tubing, excepting only for a few minutes when fresh tubes are being added. It will be seen that the cast-iron well is in this case the ultimate abutment against which the pressure is exerted in forcing the tubes down, instead of the weight of the boring machine with stones and pig iron added, as in the case where the screw-jacks are used : the hydraulic method was designed specially for the work at Gosport, and has acted most perfectly. Both the cast-iron well and the bore-hole are entirely shut off from all percolation of sea-water, by first filling up the well 30 feet with clay round the tubes, and making the tubes themselves water-tight at the joints at the time of putting them together.

In the event of any accident occurring to the tubes while they are being forced down the bore-hole, such as requires them to be drawn up again out of the hole, the prong grapnel, Fig. 212, is employed for the purpose, having three expanding hooked prongs, which slide down readily inside the tube, and spring open on reaching the bottom ; the hooks then project underneath the edge of the tube, which is thus raised on hauling up the grapnel. In case the tubes get disjointed and become crooked during the process of tubing, the long straightening plug, Fig. 213, consisting of a stout piece of timber faced with wrought-iron strips is lowered down inside them ; above this is a heavy cast-iron block, the weight of which forces the plug past the part where the tubes have got displaced, and thereby straightens them again.

Although there are many localities where the geological formation is favourable to the yield of pure water, if a boring be carried deep enough, yet it rarely happens that free-flowing wells such as those in Paris and Hull are the result. Generally after the water-bearing strata have been pierced, the level to which the water will rise is at some depth below the surface of the ground ; and only by the aid of pumps can the desired supply be brought to the surface. Various pumping arrangements have therefore been adopted to suit the different conditions that are met with.

It is not the object of the present work to treat of the forms and fittings of pumps, and the following details are only given as completing Mather and Platt's system.

It is always desirable to sink a cast-iron well, such as that at the Horse Fort, as nearly as possible down to the level at which the water stands in the bore-hole. The sinking of such a well is rendered an easy and rapid operation, with the aid of the boring machine in winding out the material from the bottom, and keeping the sinkers dry by the use of the dip bucket, shown in Figs. 223 to 225, which will lift from 50 to 100 gallons of water a minute, for taking off the surface drainage. A well having thus been made down to the level of the water in the bore-hole, the permanent pumps are then applied to the bore-hole as follows, the size of the pumps varying according to the diameter of the bore-hole. Taking the case of a 15-inch bore-hole, a pump barrel consisting of a plain cast-iron cylinder, say 12 inches diameter and 12 feet long, as shown in

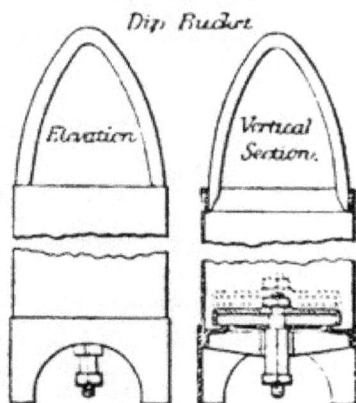

Fig. 223.   Fig. 224.

section in Fig. 228, is attached at the bottom of cast-iron or copper pipes, which are $\frac{1}{4}$ inch larger in diameter than the pump barrel, and are coupled together in lengths by flanges, Fig. 226. By adding the requisite number of lengths of pipe at the top, the pump

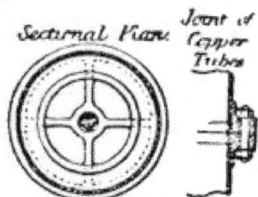

Fig. 225.   Fig. 226.

barrel is lowered to any desired depth down the bore-hole : the nearer to the depth of the water-bearing strata the better. The topmost length of pipe has a broad flange at its upper end, which rests upon a preparation made to receive it on the cast-iron bottom of the well, as at C in Fig. 228.

A pump bucket D, Fig. 228, with a water passage through it and a clack on the top side, is then lowered into the barrel, being suspended by a solid wrought-iron pump-rod E, which is made up of lengths of 30 feet coupled together by right-and-left-hand screw-couplings, as in Fig. 227. A second bucket F

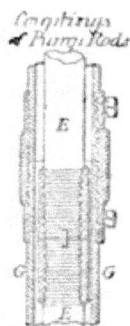

Fig. 227.

Pumping Engine and Borehole.

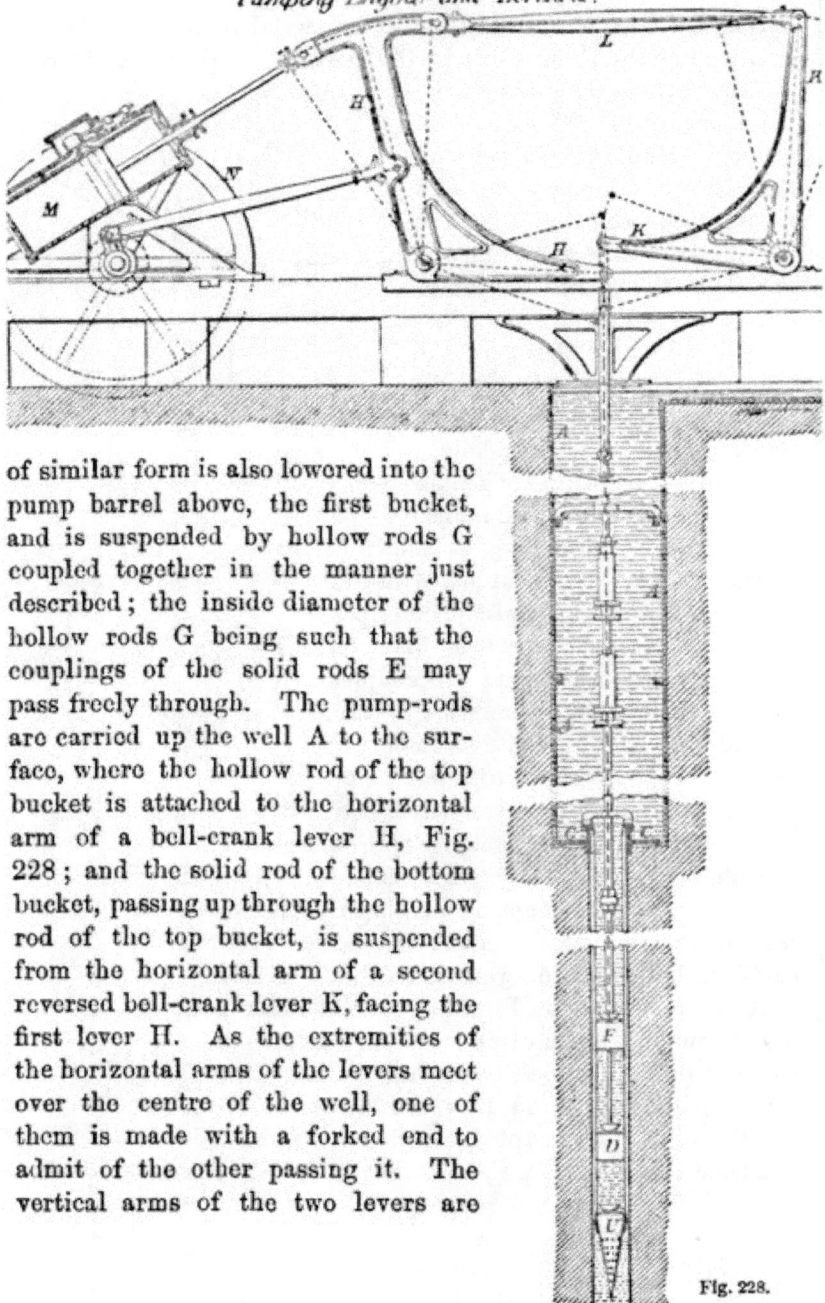

of similar form is also lowered into the
pump barrel above, the first bucket,
and is suspended by hollow rods G
coupled together in the manner just
described; the inside diameter of the
hollow rods G being such that the
couplings of the solid rods E may
pass freely through. The pump-rods
are carried up the well A to the surface, where the hollow rod of the top
bucket is attached to the horizontal
arm of a bell-crank lever H, Fig.
228; and the solid rod of the bottom
bucket, passing up through the hollow
rod of the top bucket, is suspended
from the horizontal arm of a second
reversed bell-crank lever K, facing the
first lever H. As the extremities of
the horizontal arms of the levers meet
over the centre of the well, one of
them is made with a forked end to
admit of the other passing it. The
vertical arms of the two levers are

Fig. 228.

coupled by a connecting rod L, and a reciprocating motion is
given to them by means of an oscillating steam cylinder M, the
piston rod of which is attached direct to the extremity of one of
the vertical arms; a crank and flywheel N are also connected to
the levers, for controlling the motion at the ends of the stroke.
With the proportion shown in the Figure of 3 to 4 between the
horizontal and vertical arms of the bell-crank levers, the stroke
of 5 feet 4 inches of the steam piston gives 4 feet stroke of the
pump. The reciprocating motion of the reversed bell-crank
levers causes the two buckets to move always in opposite directions, so that they meet and separate at each stroke of the
engine. A continuous flow of water is the result, for when the
top bucket is descending, the bottom bucket is rising and de-
livering its water through the top bucket; and when the top
bucket rises, it lifts the water above it while the bottom bucket
is descending, and water rises through the descending bottom
bucket to fill the space left between the two buckets. In this
way the effect of a double-acting pump is produced.

Although a continuous delivery of water is thus obtained of
equal amount in each stroke, it is found in practice that a heavy
shock is occasioned at each end of the stroke, in consequence of
both the buckets starting and stopping simultaneously, causing
the whole column of water to be stopped and put into motion
again at each stroke. As an air-vessel for keeping up the
motion of the water is inapplicable in such a situation, a modified
arrangement of the two bell-crank levers has been adopted, which
answers the purpose, causing each bucket at the commencement
of its up stroke to take the lift off the other, before the up stroke
of the latter is completed. By this means all shock is avoided,
as the first bucket gently and gradually relieves the second,
before the return stroke of the second commences.

In this improved pumping motion, which is shown in Figs.
229, 230, the two bell-crank levers H and K, working the pump
buckets, are centred one above the other, the upper one being
inverted; the vertical arms are slotted, and are both actuated
by the same crank-pin working in the slots, the revolution of
the crank thus giving an oscillating movement to the two levers

Pumping Engine
and Double-acting Pump
with Improved motion.

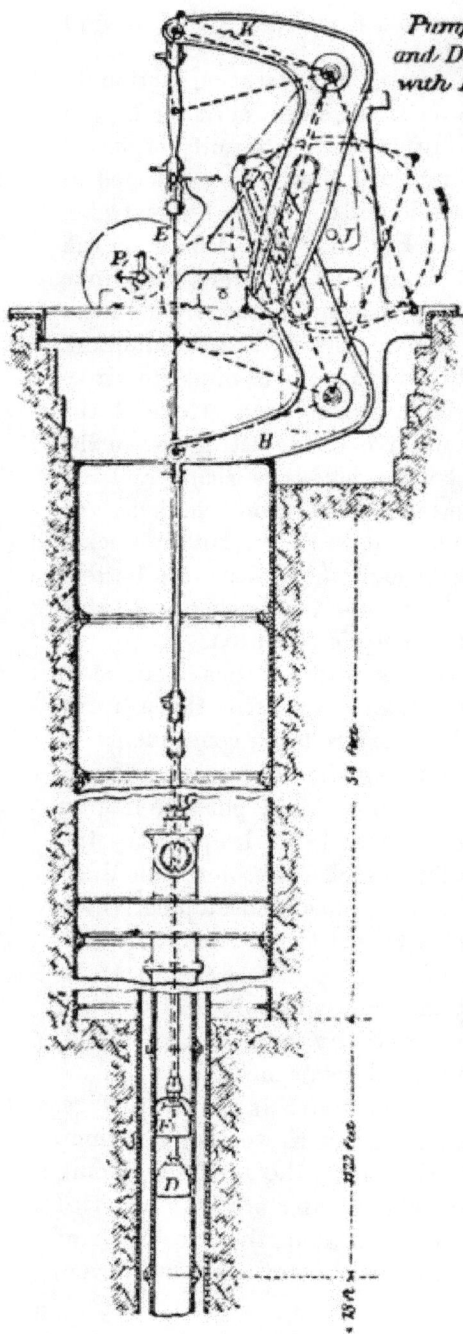

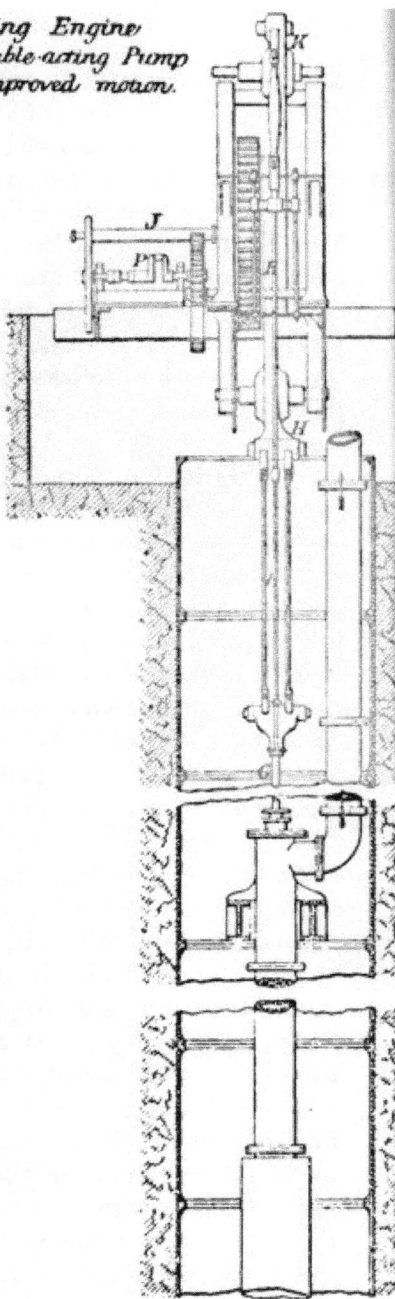

Fig. 229.

Fig. 230.

through the extent of the arcs shown by the dotted lines in Fig. 229. The solid pump rod E suspending the bottom bucket D is attached to the upper bell-crank lever K, and the hollow rod G of the top bucket is suspended from the lower lever H ; the crank-shaft J working the levers is made to revolve in the direction shown by the arrow in Fig. 229, by means of gearing driven by the horizontal steam engine P.

The result of this arrangement is, that in the revolution of the crank the dead point of one of the levers is passed before that of the other is reached ; so that the bucket which first comes to rest at the end of its stroke, is started into motion again before the second bucket comes to rest. Thus in the lifting stroke of the bottom bucket worked by the upper lever K, the bucket in ascending has only reached the position shown at D in Fig. 229, at the moment when the top bucket, worked by the lower lever H, arrives at the bottom extremity of its stroke, and the bottom bucket D, which is still rising, continues to lift until it reaches its highest position, by which time the top bucket has got well into motion in its up stroke, and is in its turn lifting the water.

AMERICAN ROPE-BORING SYSTEM.

The method of boring with a rope received great development in Pennsylvania, U.S., where the petroleum industry of the past thirty years has caused the prosecution of boring operations on a scale unknown elsewhere. As at present employed in the oil regions of the United States it is thoroughly worthy of attentive study. The following excellent description is mainly derived from an account in Cone and Johns' 'Petrolia,' a brief history of the Pennsylvania petroleum region.

The derrick or sheer-frame employed is a tall framework of timber, the bottom from 10 to 16 feet square and from 30 to 56 feet high. On the top is a strong framework for the reception of a pulley over which the drill rope passes. The floor of the derrick is made firm by cross sleepers covered with planks. A roof for the protection of the workmen is arranged some 10 or

12 feet above the floor, and in cold weather the sides are boarded
up. On one side of the derrick a windlass of peculiar construction called "the bull-wheel" is arranged, and on the other is a
steam engine, giving motion to a connecting rod which rocks
the lever, or working beam, and also by means of a belt to the
bull-wheel. The arrangement indeed, very much resembles
that of the boring sheer-frame in the frontispiece, if the windlass
were detached, and with the lever arranged to be worked by
power.

The first thing in order, is to drive the iron driving-pipe
from 6 to 75 feet, generally from 20 to 50 feet. This pipe acts
as a conductor, and prevents earth or stones from falling into
the hole while the drilling is going on. The driving-pipe in
general use is of cast iron, 6 to 8 inches in diameter, and
1 inch in thickness, in lengths of 9 or 10 feet. The driving
of this pipe is a work of difficulty, requiring the utmost skill,
since the pipe must be forced down through all obstructions to
a great depth, while it is kept perfectly vertical. The slightest
deflection from a straight line ruins the well, as the pipe acts
as a conductor for the drilling tools.

The process of driving is simple but effective. Two slide-
ways made of plank are erected in the centre of the derrick to
the height of 20 or more feet, 12 to 14 inches apart, with edges
in toward each other, and the whole made secure and plumb.
Two wooden clamps or followers are made to fit round the pipe,
and slide up and down on the edges of the ways. The pipe is
erected on end between the ways and held perpendicular by
these clamps, and a driving-cap of iron fitted to the top. A ram
is then suspended between the ways, so arranged as to drop perpendicularly upon the end of the pipe. The ram is of timber,
6 to 8 feet long, and 12 to 14 inches square, banded with iron
at the lower or battering end, with a hook in the upper end to
receive a rope. When the whole is in position, a rope is
attached to the hook in the upper end, passed over the pulley
of the derrick, down to and round the shaft of the bull-wheel.
Everything is now in readiness to drive the pipe. The belt
being adjusted connecting the engine and band-wheel, and the

rope connecting the band-wheel and bull-wheel, called the bull-wheel rop·, the machinery is put in motion, one man standing behind the bull-wheel shaft, grasping the rope attached to the ram, and coiled round the bull-wheel shaft, holds it fast, and takes up the slack in his hands, thus raising the ram to its required elevation, when it is let fall upon the pipe, which by repeated blows is driven to the requisite depth. When one joint of pipe is driven another is placed upon it, and the two ends secured by a strong iron band, and the process continued as before. The pipe has to be cleaned out frequently, both by drilling and sand pumping, or working the shell. Where obstacles such as boulders are met with, the centre-bit is put into requisition, and a hole, two thirds the diameter of the pipe, is drilled. The pipe is then driven down, the edges of the obstacle being broken by the force applied, the fragments falling into the hollow created by the passage of the bit. When this cannot be done, the whole machinery and derrick is moved sufficiently to admit of the driving a new set of pipes, or the hole abandoned. It sometimes happens that the pipe is broken, or diverted from its vertical course by some obstacle. The whole string of pipe driven, has to be drawn up again, or cut out in the manner described, p. 93, and the work commenced anew. If this is not possible, a new location is sought.

After the pipe is driven, the work of drilling is commenced. The drilling rope which is generally $1\frac{1}{4}$ inch hawser-laid cable, of the required length, from 500 to 1000 feet, is coiled round the shaft of the bull-wheel, the outer end passing over the pulley on the top of the derrick down to the tools, and attached to them by a rope socket, Fig. 231. The tools consist of the centre-bit or chisel, auger-stem or drill-bar, jars, sinker-bars and rope-socket, which are shown arranged for work in the order detailed, Fig. 237*. When connected, these

Fig. 231.

are from 30 to 40 feet in length, and sometimes more, weighing from 800 to 1600 lb., according to depth required. The process of drilling, until the whole length of the tools are on, and suspended by the cable, is slow. When the depth required to suspend the tools is obtained below the surface, the attachment between the working beam and drilling cable is made by means of a temper screw suspended from the end of the working beam, and attached to the rope by a clamp. The temper screw, Fig. 232, is from 2 to 3 feet in length, made with a coarse thread, and works in a narrow iron frame, with a nut at the lower end of the screw for the driller to let out the same as required. As the drill sinks down into the rock, the screw is let down by a slight turn of the nut by the driller, some allowing a full revolution every few blows of the bit, others once only in a few minutes, depending upon the hardness of the rock being drilled through.

Figs. 233 to 237 are other tools connected with the system. Fig. 233 is a pair of jars;

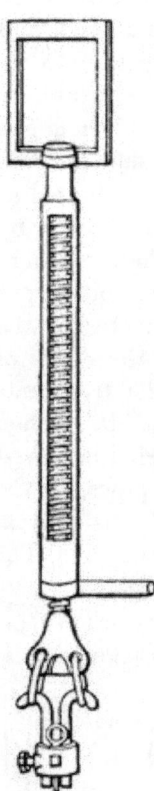

Fig. 232.

Fig. 233.

Fig. 231.

Fig. 234 a lazy-tongs for the recovery of broken ropes; Fig. 235 valve socket or catch-all; Fig. 236 a flatkey; Fig. 237 pipe clamps.

The jars, Fig. 233, attached to the auger-stem, play a highly important part in the work of drilling. They are two long links or loops of iron or steel, sliding in each other. Drillers always have about from 4 to 6 inches play to the jars, which they call the jar, and by this they can tell when to let down the temper-screw.

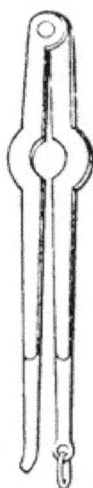
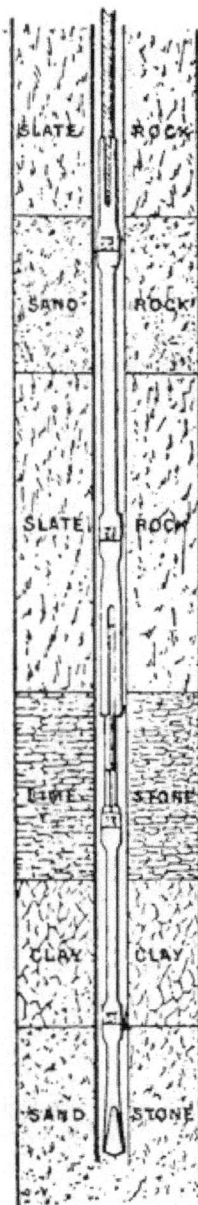

Fig. 235.   Fig. 236.   Fig. 237.

With the downward motion the upper jar slides several inches into the lower one; on the upward motion this is brought up, bringing the end of the jars together with a blow like that of a heavy hammer on an anvil, making a perceptible jar. Experienced drillers can, as soon as they take hold of the rope, tell how much "jar" they have on.

Fig. 237*.

In drilling, the tools are alternately lifted and dropped by the action of the working beam on its rocking motion. One man is required constantly in the derrick, to turn the tools as they rise and fall, to prevent them from becoming wedged fast, and to let out the temper-screw as required. This is one of the most important duties of the work, requiring constant attention to keep the hole round and smooth. The centre-bit or chisel is run down the full length of the temper-screw; it is about $3\frac{1}{2}$ feet in length, with a shaft $2\frac{1}{4}$ inches in diameter, and a cutting edge of steel $3\frac{1}{2}$ to 4 inches in width, with a thread on the upper end by which it is screwed on the end of the auger-stem. The reamer is about $2\frac{1}{2}$ feet in length, having a blunt instead of a cutting edge, with a shank $2\frac{1}{4}$ inches in diameter, terminating in a blunt extremity $3\frac{1}{2}$ to $4\frac{1}{2}$ inches in width by 2 inches in thickness, faced with steel. The weight of heavy centre-bits and reamers average from 50 to 75 lb. each.

The centre-bit is followed by the reamer, to enlarge the hole to make it smooth and round. The sediment, or battered rock, is taken out after each centre-bit, and again after every reamer, by means of a sand pump let down in the well for the purpose. The sand pump now in use is a cylinder of wrought iron, 6 to 8 feet in length, with a valve at the bottom, and a strap at the top, to which a $\frac{1}{2}$-inch rope is attached, passing over a pulley suspended in the derrick some 20 feet above the floor, and back to the sand pump reel attached to the jack frame, and coiled upon the reel-shaft.

This shaft is propelled by means of a friction pulley, controlled by the driller in the derrick, by a rope attached. The sand pump is usually about 3 inches in diameter. Some drillers use two, one after the centre-bit, and a larger one after the reamer, the two being preferable. When the sand pump is lowered to a requisite depth, it is filled by a churning process of the rope in the hands of the driller, and is then drawn up and emptied. This operation is repeated each time the tools are withdrawn from the well, the pump being let down a sufficient number of times to remove the drillings. The fall of the tools is from 2 to 3 feet. This labour goes on, first tools and then sand pump, until the well is drilled to the required depth. Abundance of water is found in the wells, both for rope and tools, from the commencement. It flows in from the surface veins, and from the larger ones below.

The following are practical directions in employing the rope-boring system.

The driller takes his seat on a high stool above the chosen spot, adjusts the drill with great care, and through the conductor-pipe, striking from thirty to forty blows a minute.

Between the strokes the tools require to be moved round. With this is also continued a slight downward motion every few strokes, by a turn of the temper screw.

The drill is kept moving up and down, cutting from 1 to 6 inches and even 12 inches of rock and shale an hour according to hardness. At intervals the centre-bit is drawn up, badly worn and battered, and a reamer let down to enlarge the hole and make it smooth and round, and these are followed by the sand pump.

The first few hundred feet are generally gone through without difficulty, provided all the arrangements have been made with care at the beginning, and the drillers are skilful. Difficulties occur farther down that test to its utmost endurance the most persistent energy.

Sometimes they are attributable to a want of caution on the part of the driller, from imperfection in the material of, or improper dressing, or tempering the drill, but more often to circumstances unforeseen and unavoidable. In its passage the drill not unfrequently dislodges gravel or fragments of hard rock, that have a tendency to, and often do wedge it fast in the hole, from which it is only dislodged by the most persistent "jarring."

The reamer is also subject to the same mishap, or a sand pump breaks loose from its rope, and has to be fished up. When the bit or reamer becomes so firmly imbedded as to render its removal impossible by jarring or breaking it in pieces, the well is abandoned.

N

Sometimes a bit or reamer breaks, leaving a piece of hard steel securely in the rock several hundred feet below the surface. Where the fragment is small, it is pounded into the sides of the well, and causes no farther annoyance. When it is larger the difficulty is greater, and not unfrequently insurmountable. The bit or reamer sometimes becomes detached from the auger-stem, by the loosening of the screw from its socket. This difficulty is often greatly heightened from the fact that the workman may not be aware of its displacement, and for an hour or two be pounding on the top of it with the heavy auger-stem. Various plans are resorted to to extract the fastened tool, and a large number of implements have been devised for fishing up the same. The first instrument used is an iron with a thin cutting edge, straight, circular or semicircular, acting as a spear, or to cut loose the accumulations round the top and along the sides of the refractory bit or reamer, so as to admit a spring socket that is lowered by means of the auger-stem over the top of it, and lays hold upon the protuberance just below the thread.

If the socket can be made fast, the power of the bull-wheel and engine is brought into requisition, and in a great number of cases it is brought to the surface. In the jarring and other operations rendered necessary in cases of this kind, the entire set of tools, 40 to 60 feet in length, may become fastened, and cases are of frequent occurrence where two and even three sets of tools have become fastened in a well, as they were successively let down to extricate the first ones. The difficulty described is liable to occur at any stage of the work, and its frequency increases with the depth.

In addition to the difficulties mentioned, there is yet another, far more dreaded by the driller. This is what is called a mud vein. It is a thin stratum of mud or clay, from one to several inches in thickness, generally met with at the depth of from 400 to 900 feet. Mud veins abound in most of the producing localities and not a few operators regard them as invariably indicating an abundant supply of oil.

This mud or clay is of a most tenacious character, is highly annoying to the operator when drilling, and in many cases disastrous. Though not deemed of much importance as an obstacle in the beginning of the development, the mud vein exhibits new features in different localities. The mud suddenly flows into the well while the process of drilling is going on, settling round the drill, bedding it as firmly almost as the rock itself. Its presence is often indicated to the driller by the sudden downward pressure on his rope. When drilling on or below it, the workman when about to withdraw his drill, will have assistance at the bull-wheel, and the instant the working beam ceases its motion, a few turns will be taken on the wheel, so as to raise the bit above the mud, as it sets almost as quickly as plaster of Paris; sometimes this mud will flow into the hole for a depth of twenty or more feet, burying as it were, the entire drilling tools and attachments. This renders the jars useless. By attaching a cutting instrument to rods, the rope above the sinker-bar is cut, and then a spear-pointed instrument substituted, with which, by means of a light set of tools, the substance round the tools is forced from them, an extra pair of jars lowered, and efforts made to jar the tools loose.

The spear is sometimes shaped like a common wedge, faced with steel at the cutting edge, made thin. A half-circular instrument, made in similar manner, is also used. The mud socket, circular shaped with thin edge, terminating on the inside with an abrupt shoulder corresponds with the ordinary sheel or clay auger, and is used in a similar manner.

A large number of appliances have been invented for the dislodgment of fastened tools, but many of these are very complicated. The main thing sought to have is an instrument that in the first place will remove the material round the top of the fastened implements, to be followed by others acting on the principle of a clamp sufficiently powerful to retain its hold and allow the jarring of the tools loose, or the drawing of them up.

One most effective instrument for the dislodgment of tools is in use. This consists of a number of heavy iron rods or bars, similar to an auger-stem, weighing from 10 to 11 tons. It can be made of any desired length or weight. It is lowered over the head of the tools, and these screwed fast into a suitable socket arranged at the ends of the rods, and worked from the top. When a set of tools are fast, each separate piece is unscrewed, the apparatus acting as a left-handed screw. Each piece, as loosened, is brought to the surface. This is stated to be the most efficient device yet invented, and is in extensive use. By applying the full force of the engine, these 2½ inch iron rods are frequently twisted like an auger. They are lowered and raised from the top by jack screws.

It will be seen that the system has many features in common with European practice. The centre-bit and reamers are but other names for variously shaped chisels, whilst the jars serve a similar purpose to that of the sliding joints illustrated at pp. 134 and 136. As a cheap method of putting down deep bore-holes through shales, limestones, and soft rocks it is very useful, but it must certainly be supplemented by others when hard or troublesome beds are met with.

THE DIAMOND DRILL.

The diamond drill can be employed with advantage in boring for water, particularly where hard rock has to be dealt with. It depends for its action upon abrasion; a number of diamonds are set in a steel crown, Fig. 245, which is attached to hollow rods, Figs. 240 to 244, and rotated at from 40 to 300 revolutions a minute under pressure varying with the nature of the rock, from 300 to 800 lb. being applied with small holes, rising to as much as 1100 lb. for larger ones.

The diamonds employed are a variety which is found massive in small black pebbles called "carbonardo" or carbonate, having a specific gravity $3 \cdot 102$ to $3 \cdot 416$; they are pure carbon excepting $2 \cdot 07$ to $2 \cdot 27$ per cent.

The boring machine consists of two vertical girders, Figs. 238,

239, carrying between them bearings which support a hollow stem A of sufficient size to grasp and turn the boring bars. This stem has a rotary motion imparted to it by means of an

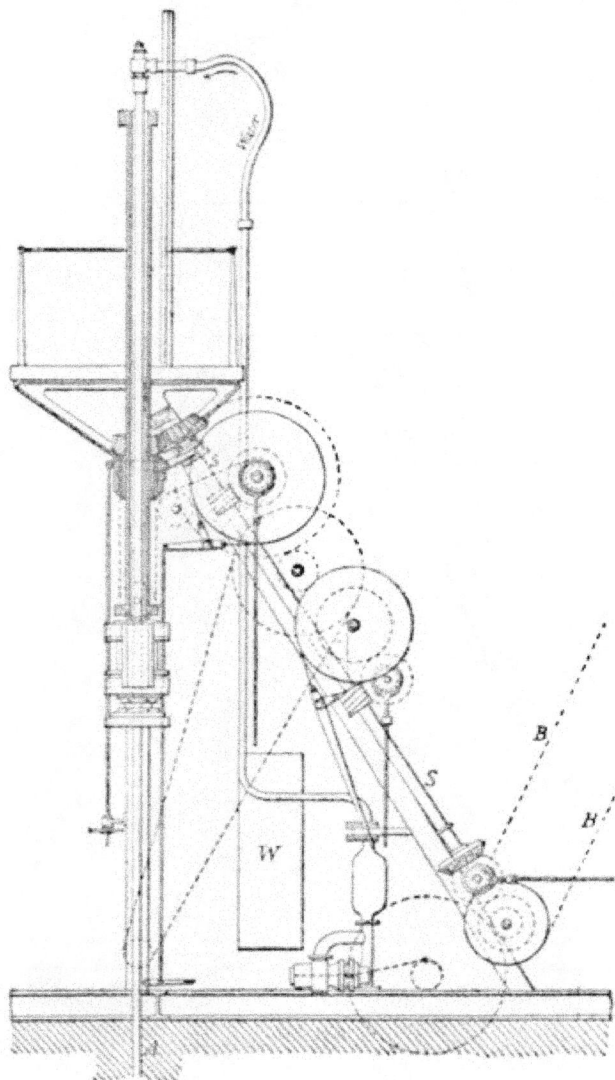

Fig. 238.

inclined shaft S driven by bevel gear, power being transmitted
by means of a belt B from the fly-wheel of a portable engine.

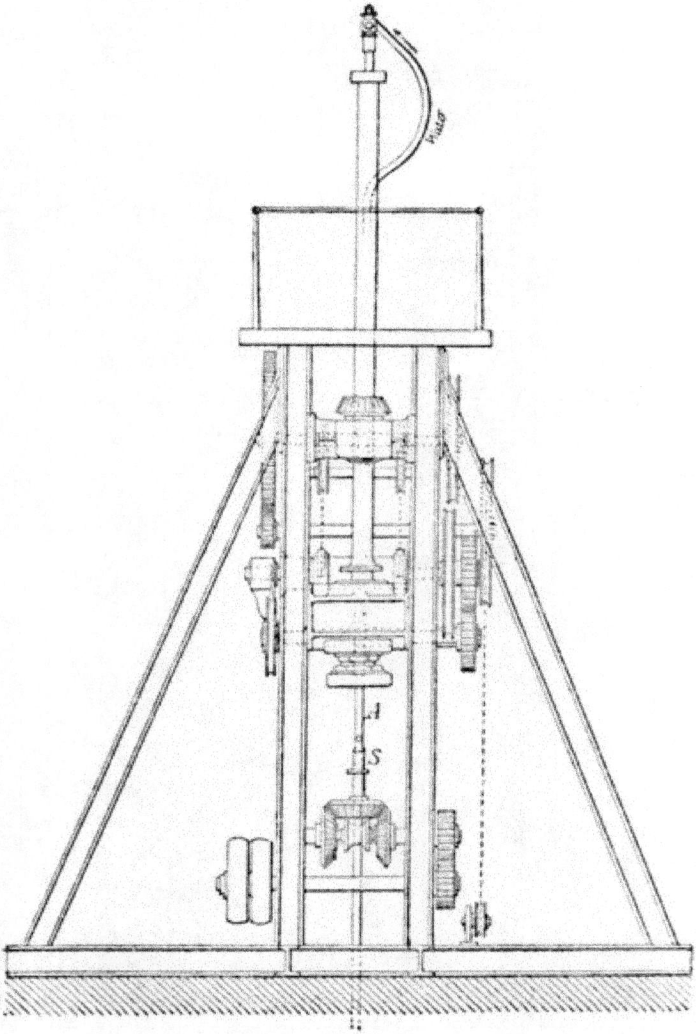

Fig. 239.

Fig. 245 is of the older form of crown, and Figs. 240, 241,
243, 244, give details of the hollow rods and their connections.

Fig. 244 is a section of a bore-hole showing the core in the interior of the rods. Fig. 242 is the extractor which is substituted for the diamond crown when the core is to be broken off.

The details of the diamond drill have been subjected to much improvement by J. R. Gulland, and in two important borings

Fig. 240. Fig. 241. Fig. 242.

 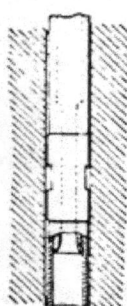 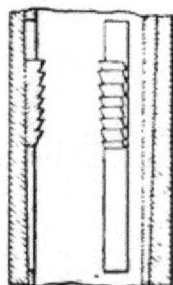

 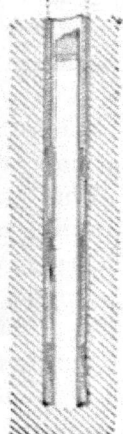 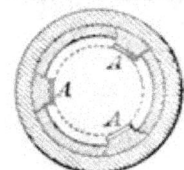 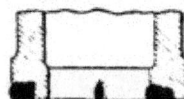 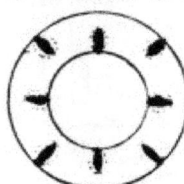

Fig. 243. Fig. 244. Fig. 245.

described by H. J. Eunson, in the Proceedings Inst. C.E. 1883, from which the following account is taken, Gulland's machine was the one used. Here the largest crown was 23 inches in external diameter, and contained fifty stones, having an aggregate weight of more than 300 carats. The crown,

Fig. 216.

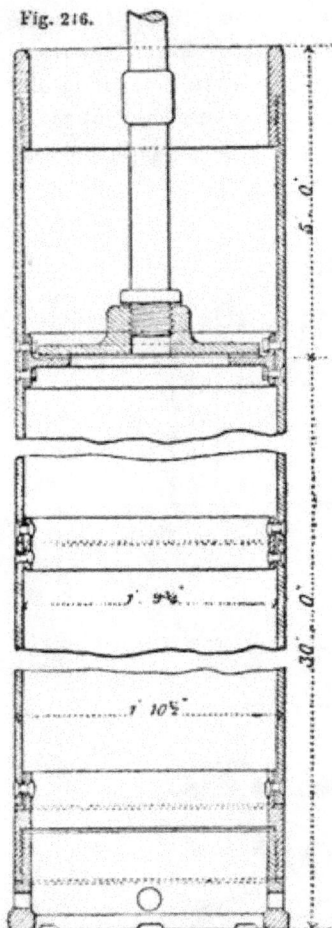

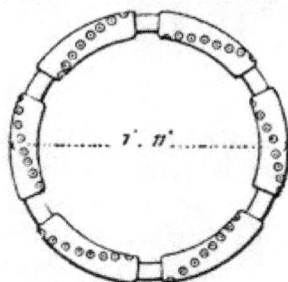

Fig. 247.

Figs. 246, 247, is of improved construction, is screwed to the core tube, which serves to keep the drilling vertical, and contains the core as it is drilled. In the first size the core tube was $22\frac{1}{2}$ inches external diameter, 30 feet in length, and of wrought iron. The rods connecting the core tube with the surface machinery fit into a plate at the top of the tube, above which is a 5-feet length of tube of the same diameter, and open at the top. This receives the coarser particles falling from the water flowing upwards after washing away the *débris* in drilling, also any fragments which may be detached from the sides, thus preventing the crown from becoming clogged in the bore-hole.

The boring rods are tubes of drawn steel, $3\frac{1}{2}$ inches outside diameter, and $\frac{3}{8}$ inch in thickness, in lengths of 5 feet, connected by steel collars.

During the operation of boring, a continuous supply of water is pumped down through the hollow bore rods, to keep the crown cool and carry off the *débris* formed by the erosion of the strata by the crown. The water flows through channels cut in the face of the crown, rises on the outside of the core tube to the surface, and is collected in settling-ponds, where the sediment is deposited. About 3500 gallons of water an hour were required, the water, after settling, being used over again.

The boring machinery on the surface was similar to the arrangement, Figs. 238 and 239, and consisted of a strong framework of wrought iron, having two principal pillars in front, one of cast iron, which forms the chief support of the upper part of the machine, which is also stayed by raking supports from behind; the other, a circular upright shaft, running in a shoe at the bottom, and a bearing at the top. This is the main shaft for transmitting the power from the machinery to the rods by a system of bevel wheels.

The crosshead from which the rods are immediately driven slides on the circular shaft, and the motion to the rods is given by a wheel fixed to the crosshead, which works on the shaft by means of a feather key. The crosshead is thus enabled to slide on the vertical shaft and follow the rods as they sink in boring. At the back of the two upright pillars, and between the raking stays, are the different parts of the hoisting apparatus for drawing and lowering the tools. When drilling, a load W, Fig. 238, is attached to counterbalance the weight of rods as the boring becomes deeper, the pressure on the crown being kept constant at about 10 cwt.

The rods, on account of the height of the sheer-legs, could be raised and lowered in lengths of 40 feet; the men could raise a length of 40 feet by the machine, disconnect it, and lay it down in front of the machine in three-and-a-half minutes; the reverse operation, that of picking one up, connecting it, and lowering it in the hole, could be accomplished in two-and-a-half minutes.

The machine was worked by a 20-HP. portable engine, but 40-HP. were frequently indicated, and the drilling machine, weighing upwards of 20 tons, was repeatedly rocked to and fro under the great strain which had to be exerted in freeing tools which had become fast in the bore-hole, when the drill worked unevenly on account of a small stone or other impediment under the face of the crown, or when an extra force was necessary to break off the core. The crown was revolved at first at about forty revolutions a minute, and this speed was increased to as many as one hundred and fifty when in favourable strata. On a depth of 5 feet having been bored, the crosshead was disconnected from the rods, raised to its full height, and a 5-feet rod inserted.

After a length of core had been drilled, sometimes nearly 30 feet, which was the limit that the core tube would contain, the tube and crown were drawn to the surface, the crown was unscrewed, and the extractor, Figs. 248, 249, fixed in its place. This tool consists of an annular ring of steel, 9 inches in depth, from the sides of which turning inwards are steel clutches or teeth, it is lowered over the column of core left standing in the hole, the projecting teeth laying hold of the core. The crosshead is next connected, and the core pulled asunder and drawn in the tube to the surface, and upon unscrewing the extractor the core can be removed. The hole is thus left clear and ready for the next drilling.

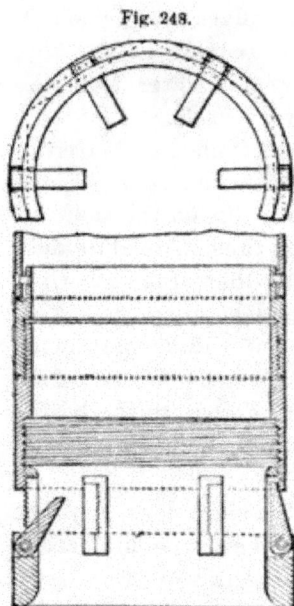

Fig. 248.

Scale

Fig. 249.

Frequently the core, or part of it would be broken off and become fixed in the core tube, coming up with the crown when first drawn. A ring of steel fitting inside the core-tube, and which clipped the core as it was drilled, was tried as an extractor, but owing to the comparatively soft nature of the clay it failed to grip it sufficiently. This ring was not tried in the harder strata.

In the larger sizes nearly the whole of the core drilled was extracted, though in some cases the clay was washed away by the water pumped through the rods; or if the core became broken, the two surfaces would be worn by the broken piece revolving with the tube, which was particularly noticed when the hard and soft beds alternated in quick succession, and when the core was sandstone large quantities were thus lost.

Table I., p. 188, gives the statistics of the boring through the different strata at Kettering Road, with the several sizes of crowns; while Table II. contains the results obtained from a boring at Gayton, 5 miles south-west of Northampton.

The progress of the boring at Northampton was much hindered by accidents and delays. The most numerous of these were caused by the breakage of the rods, the place of fracture usually being the collar, though in some cases the thread of the rods was stripped. When boring in the clay seven collars were broken, and in the quartzite five; five more in clearing out sediment, on account of the fragments of the rock and small stones which had fallen from the sides; and two were also broken when extracting the core. To make the connection again when such an accident happened, the upper length of rods was drawn to the surface, and a bell tap, Fig. 251, attached, lowered, and revolved over the rods left standing in the hole; a screw was thus cut on the broken ends of the rods, to which the tap was firmly attached, and the whole of the rods drawn to the surface, and the broken collar replaced by a new one. Two kinds of taps were used for this purpose; one a bell tap for lowering over the rods, the other a taper tap, Fig. 250, for insertion in the hollow of the rods. Upon one occasion, in recovering the rods which had been broken, a second collar broke, the tap at the time being in the hole; fortunately this second breakage occurred in the 24-inch pipes and in this case the water was withdrawn and a man lowered who made the connection. The time occupied in repairing such accidents varied from an hour or two, to sometimes more than a day.

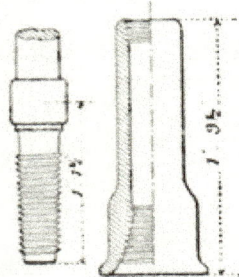

Fig. 250.   Fig. 251.

Several times during the operation of boring small pieces of

Table I.—Boring at Kettering Road, Northampton.

Diameter of Crown.	Depth Drilled.	Number of Days Drilling and Extracting.	Average Depth a Day.		Nature of the Strata.	Diameter of Core.	Quantity of Material Extracted.
inches	feet		ft.	in.		inches.	per cent.
23	77	17	4	6½	Lias clay	19¼	
20½	97	15	6	5½	„	16¾	
18	106	16	6	7½	„	14½	
15¾	55	11	5	0	„	12¼	
„	68	10	6	9	Sandstones and marls	„	95
„	25	15	1	8	Quartzite	„	100
„	20	5	4	0	Limestone and shale	„	98

Table II.—Boring at Gayton, South-West of Northampton.

Diameter of Crown.	Depth Drilled.	Number of Days Drilling and Extracting.	Number of Hours Drilling.	Average Depth. A Day.		Average Depth. An Hour.		Nature of the Strata.	Diameter of Core.	Quantity of Material Extracted.
inches	feet			ft.	in.	ft.	in.		inches	per cent.
18	125	11	104	11	4	1	3	Lias clay	14½	88
15¾	148	13	127	11	4½	1	2	„	12¼	90
13⅝	182	17	183	10	8½	1	0	„	10¾	92
11⅞	117	10	100	11	8	1	2	„	9¼	88
„	63	8	60	8	0	1	0½	Red marl and sandstone	„	64
10⅛	215	25	213	8	7	1	0	Lower carboniferous Limestone and shalo	7¾	84
								Sandstones	„	68

iron getting into the hole necessitated the stoppage of work to extract them. These pieces had broken from the top of the 22½-inch cast-iron lining tubes. In lowering the tubes they broke away from the bayonet joint and fell a distance of 480 feet to their position at the bottom of the hole. The tubes, which weighed upwards of 10 tons, passed through a bed of 50 feet of accumulated sediment. In raising and lowering the tools they caught against the ragged top of the tube where it had broken from the bayonet joint, and thus fragments were broken off which fell to the bottom. These pieces of iron, when drilling in the clay, were, in the majority of cases, forced into the sides of the bore-hole; but in drawing the 17-inch tubes to enlarge the hole, the sides falling in carried the fragments of iron with them, which in the harder beds became a source of great trouble. In the clay a wedge-shaped tool was used to cut a hollow in the bottom of the hole, and at the same time sweep the small pieces of iron into it; the crown was then lowered and a length of core drilled and extracted, and the iron brought to the surface in the hollow on the top of the core. In the harder beds, where this tool would not work, a heavy chisel was used, and the hole was jumped, the sediment, with the iron, being extracted by a shell. Some of the iron was extracted by a plug of wood, which fitted into the core tube, being forced several times upon the bottom, causing numerous pieces to adhere to it.

The core tube sometimes became clogged in the hole; in this case the lifting chain was removed from the framework and replaced by a hemp rope 6 inches in diameter, reefed four times through blocks and attached to the windlass. By using this rope and blocks a more elastic and greater strain could be exerted; but the rope was more than once broken before the core tube could be moved.

For prospecting purposes and for holes of small diameter the diamond drill is arranged in a more compact form than in Fig. 238. A drill of this class is made by the Diamond Drill Company of Pennsylvania, where the engine is so arranged that its motion is transmitted direct to the bevel gears turning the rods of the machine; although small it is very effective.

Another modification which has come under the writer's direct notice is that devised by Olaf Terp. The Diamond Rock Drill is useless in soft clays, and nearly so in loose gravels, and thin beds of this character very seriously retard the progress of a boring. To obviate this, Terp has invented a steel borer head of peculiar construction, which is substituted for the diamond crown when a soft stratum is met with. The hollow rods terminate in the centre of the borer head in the form of a nozzle, through which water under pressure is injected, and so forces up to the surface, as mud, the material displaced at the bottom of the bore-hole. When rock is again reached the diamond crown is replaced.

CHAPTER VIII.

EXAMPLES OF WELLS EXECUTED, AND OF DISTRICTS SUPPLIED BY WELLS.

PERMIAN STRATA.

Durham.—Large quantities of water are pumped from the lower Permian sandstone beneath the magnesian limestone of this county, and are used for the supply of the towns of Sunderland, South Shields, Jarrow, and many villages. The quantity, calculated by Daglish and Foster to reach 5 millions of gallons a day, is obtained from an area of 50 square miles overlying the coal measures. The water level has not been lowered in the rock by these operations. Along the coast it is that of mean tide, and inland rises to a level of 180 feet. In the coal measures below there is little water, and that little is saline. Sedgwick gives the strata as red gypseous marls, 100 feet; thin bedded grey limestone, 80 feet; red gypseous marls, slightly salt, 200 feet; magnesian limestone, 500 feet; marl slate, 60 feet; lower red sandstone, 200 feet.

Coventry.—Warwickshire. The town is supplied with 750,000 gallons of water a day from two bore-holes made in the bottom of the reservoir. The bore-holes are respectively 6 inches and 8 inches diameter, and 200 feet and 300 feet deep. The town is situated on the Permian formation, but Latham states that the supply is procured from the red sandstone, and, from observations made, it has been found that the two bore-holes yield water at the rate of 700 gallons a minute.

TRIAS STRATA.

Birkenhead.—There are here several deep wells belonging to the Tranmere Local Board, the Birkenhead Commissioners, and

the Wirral Water Company, yielding together about 4,000,000
gallons a day. Figs. 252, 253, show a section and plan of the
No. 2 or new engine-well at the Birkenhead Waterworks.
The shaft is 7 feet diameter for 105 feet, with a bore-hole 26
inches for 35 feet, 18 inches for 16 feet, 12 inches for 99 feet,
and 7 inches for 150 feet, or a total depth from surface of 405
feet. The water level is about 95 feet from surface when the
engine is not at work. At the upper water level shown in the
26-inch hole, the yield was at the rate of 1,807,400 gallons in
twenty-four hours, at the lower level at the rate of 2,000,000
gallons in the same time. At the water level indicated in the
7-inch bore, water was met with in large quantities. The old
engine well is almost identical.

Figs. 254, 255, are a section and plan, and Fig. 256, enlarged
parts of the well at Aspinall's brewery, Birkenhead. It consists
of a shallow shaft 5 feet in diameter, and steined, continued by
means of iron cylinders 3 feet 3 inches in diameter and 50 feet
in depth. When sand with much water of poor quality was
met with, a series of lining tubes was introduced from the point
A A, the space between these and the cylinders being filled with
concrete. The tubes were discontinued at the sandstone, and
the lowest portion of the hole, 3 inches in diameter, is unlined.
The water overflows.

Figs. 257, 258, are a section and plan of the well at Cook's
brewery, Birkenhead. The shaft is 6 feet diameter, lined with
9-inch steining, and is 66 feet deep. At 29 feet from surface
it is enlarged for the purpose of affording increased storage
room for the water. There is a 16-inch pipe at bottom of shaft
49 feet deep, continued by a 12-inch bore-hole 13 feet into the
red sandstone. The water level is 27 feet from the surface of
the ground.

Birmingham.—Out of the 7,000,000 gallons a day supplied to
the town in 1865 by the Waterworks Company, 2,000,000 were
derived from wells in the new red sandstone. In that year an
Act was passed authorising the sinking of several new wells,
whereby the quantity has been greatly increased.

Burton-on-Trent.—Fig. 259 is a section of the well at the

London and Colonial Brewery. Extraordinary precautions were taken in constructing this well to obtain the water from the

Fig. 252.

NEW ENGINE WELL,
BIRKENHEAD WATERWORKS.

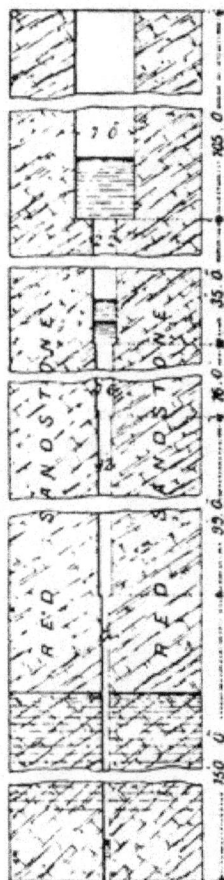

Fig. 254.

WELL AT ASPINALL'S BREWERY,
BIRKENHEAD.

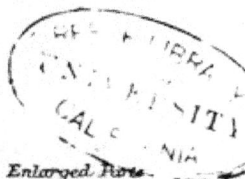

Enlarged Pipes.

at A. A

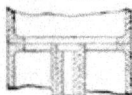

at B. B

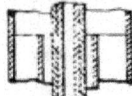

at C. C

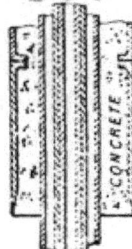

at D. D.

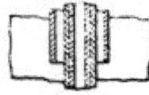

at E. E.

Fig. 256.

PLAN

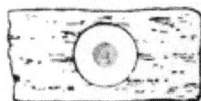

Fig. 253.

PLAN

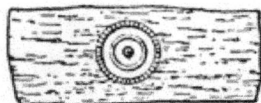

Fig. 255.

o

lower strata perfectly free from admixture with that from above. There is a steined shaft within which is an iron cylinder, and this again is lined with brick steining backed with concrete. The bore-hole, 182 feet deep and 4 inches diameter, is lined throughout with copper tubes. At the top the bore-hole is surrounded with a short tube upon which a thread is cut, so that if necessary a pipe may be screwed on and up to surface. The water rises to within 6 feet 3 inches of the level of the ground. Fig. 260 is an enlarged section of the arrangements at the top of the bore-hole, and Fig. 261 an enlarged section of the pipe joints.

Coventry.—Warwickshire. The town is constantly supplied from 4 bore-holes, one of which yields at least 750,000 gallons of water a day. Two of the bore-holes are respectively 6 inches and 8 inches diameter, and 200 feet and 300 feet deep, sunk through alternations of marl and sandstone.

Crewe.—Cheshire. A very plentiful supply of water for the requirements of the town and works of Crewe is obtained from a well sunk in the new red sandstone. The water is said to be very pure, and from the analysis of Dr. Zeidler it appears that there are only 6·10 grains of solid matter to the gallon.

Five Lane Ends, near Farnworth.—Lancashire. Well 3 feet 4 inches diameter, 86 feet deep, with bore-hole 3½ inches diameter to a depth of 170 feet from surface. The water stood at 83 feet from surface, or about 52 feet above Ordnance datum.

STRATA :—	Feet.	In.
Sandy Soil	33	0
Very fine Yellow Sandstone	7	0
Fine White Banded Sandstone, gritty in parts ..	38	0
Fine Yellow Sandstone	22	0
„ Hard Sandstone	8	0
Loamy Sandstone	9	0
Fine Sandstone, with " millet-seed " grain	7	0
Light Green and Blue Clay	2	0
Red Clay	1	0
Bright Red Sandstone	3	0
UPPER COAL MEASURES :—		
Purple Marl	1	0
Dark Red Earthy Limestone	2	0
Purple and Mottled Marl	2	0
Carried forward	135	0

Fig. 257.

Well at Cook's Brewery, Birkenhead.

Fig. 259.

Well at London and Colonial Brewery, Burton-on-Trent.

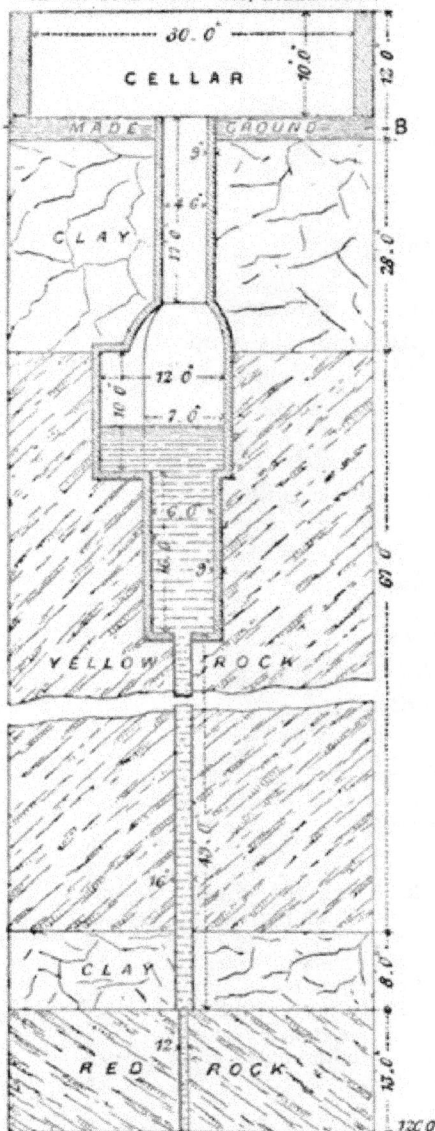

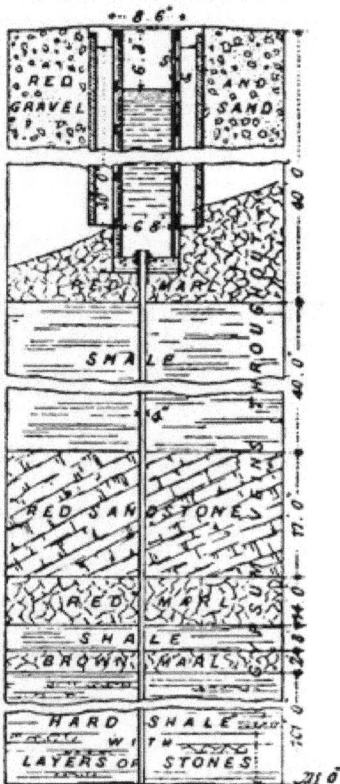

PLAN A.B.

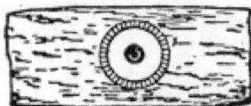

Fig. 258.

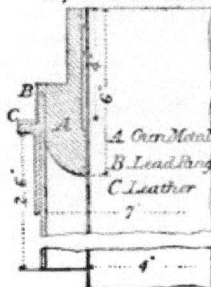

Top of Bore Hole.

A. Gun Metal
B. Lead Ring
C. Leather

Fig. 260.

Enlarged Section of Pipe Joints

Fig. 261.

	Feet.	In.
Brought forward	135	0
Calcareous Marl	3	0
Marl	3	6
Green Clay..	3	6
Red Clay	4	0
Marl	3	6
Grey Limestone..	1	6
Argillaceous Limestone	2	6
Red Marl	4	6
Marl	9	0
Total	170	0

Parkside.—Lancashire. Well on property of London and North Western Railway Company, 80 feet deep, with a borehole of 14 to 10 inches diameter to a depth of 296 feet from surface. Level of water, 69 feet from surface.

STRATA ;—

	Feet.
PEBBLE BEDS :—	
Reddish-brown and White Sandstone, with quartz pebbles	110
Coarse Brown Sandstone	4
Fine Yellow Sandstone	1
Grey Sandstone, with pebbles	4
Fine Red Sandstone	3
Grey Rock and large pebbles	3
Fine Red Sandstone	3
Fine flaggy and micaceous Yellow Sandstone ..	16
Loam, with fragments of grit	1
Reddish Loamy Sandstone	5
Red Marl	32
Fine Bright Yellow Sandstone	2
Fine Red Sandstone	1
Fine Pale Red and White Sandstone	7
Fine Brown Sandstone	8
Red Marl	4
Soft Brown Sandstone, with "millet-seed" grain ..	3
Fine Grey Sandstone, nodules of iron pyrites ..	13
Light Red Sandstone	3
Fine Brown porous Sandstone, plenty of water ..	47
Coarse Light Brown Sandstone, "millet-seed" grain	6
Concretions of "Millet-seed" Sand, cemented by iron pyrites, generally yellow or copper-coloured	2
Bright Red porous Sandstone "millet-seed" grain	12
Lumpy Ferruginous Sandstone	1
UPPER COAL MEASURES ;	
Purple and Green Mottled Marls	5
Total	296

Prescot.—Lancashire, neighbourhood of. The following table contains a list of the principal wells of the district, which draw their supply from the new red sandstone. The water is obtained from the three subdivisions of the Bunter which, though varying locally in texture, may be regarded as porous throughout. The water level in this mass of rock forms a slightly undulating plain higher inland than at the sea coast, and rising under high ground.

Locality.	Well.		Bore-hole.		Above O.D.	Pumped in 24 hours.
	Depth.	Dia-meter.	Depth.	Dia-meter.		
	feet	feet	feet	inches	feet	gallons
Dudlow Lane	247	12 × 9	196	18	198	1,240,440
Belle Vale	—	—	--	4	52	58,000
Netherlee Bridge	—	—	—	—	37	45,000
„ „	—	—	—	—	37	350,000
Cronton	—	—	—	—	65	800,000
Whiston	225	9	87	18	200	} 938,000
„ auxiliary	225	10·5	240	18	„	
Litton A	50	10	—	—	10	} 900,000
„ B	30	10	270	24	15	
Eccleston Hill	210	10	178	—	260	
Winwick A	50	—	200	—	110	
„ B	50	—	—	—	„	
Garston Ironworks	100	—	251	6	15	240,000
Dungeon Stoneworks	—	—	260	—	35	18,000
Gaskell, Deacon & Co. A ..	30	5	825	3	10	} 500,000
„ „ „ B ..	39	12	639	4	„	
„ „ „ C ..	37	8	429	9 × 6	„	
Mathieson & Co.	30	4·5	336	6	10	4,000
Sullivan & Co. A	58	6	338	4	25	140,000
„ „ B	60	10	349	14	15	600,000
Warrington Wire Co.	—	—	212	18	—	63,360
Roberts, Dale & Co.	—	—	225	9	—	28,000
Jas. Owen & Co., Winwick ..	—	—	212	18	—	461,000
Runcorn Waterworks	300	24 × 8	98	14	250	380,000

Preston Brook.—Lancashire. Well at the tan-yard, 9 feet in diameter to a depth of 51 feet from the surface, with a bore-hole to a further depth of 404 feet. The water stands at 62 feet from surface.

Section as follows ;—

GLACIAL DEPOSITS ;—	Feet.	In.
Red Clay	34	0
Sand	12	6
Stony Red Clay	136	6
RED MARLS ;—		
Red Marl	9	0
Sandstone	6	6
Red Marl	15	6
Red Sandstone	13	0
Marl	4	0
Sandstone	6	0
Marl	3	0
Sandstone Marl	199	0
Hard Sandstone	6	0
Red Marl	5	0
Sandstone	5	0
Total	455	0

This boring probably ended in the waterstones.

Winwick.—Lancashire. Well at Warrington Waterworks, 9 feet in diameter to 127 feet 5 inches, with a bore-hole of 14 inches diameter to end ;—

STRATA ;—	Feet.	In.
Fine-grained Sandstone "pebble beds"	127	0
Compact Sandstone, with large round grains throughout, and including a bed of shale	45	0
Red shale	10	5
Fine-grained Pale Red Sandstone	6	0
„ „ Grey Sandstone	2	0
Red Shale and Calcareous Sandstone	11	0
Hard Fine-grained Calcareous Red Sandstone ..	11	0
Shale	2	0
Red Sandstone, with fragments of shale, hard towards bottom	15	0
Shale	31	7
Soft Sandstone	10	5
Fine Red Sandstone	6	0
Soft Grey Sandstone, with iron pyrites	21	7
Very Soft Red Sandstone, bands of shale	31	0
Red Shale	11	0
Calcareous Green and Purple Marls	19	0
Fine-grain Red Micaceous Sandstone	5	0
Carried forward	365	0

		Feet.	In.
Brought forward	.. .. ..	365	0
Dark Green and Purple Shales	.. ..	20	0
„ Red Calcareous Marl	.. ..	11	0
„ Red Shale	.. ..	3	0
.. Calcareous Marl	.. ..	9	0
Limestone	.. ..	4	0
Total	.. ..	412	0

Leamington.—The well in this town is situated at the foot of
Newbold Hill, and is 5 feet in diameter and sunk to a depth of
50 feet. At the bottom of the well a bore-hole, part of the way
18 inches and the remainder 12 inches in diameter, is carried
down 200 feet. It passes through alternating beds of marl and
sandstone, and the surface water met with has been bricked or
puddled out. The yield is about 320,000 gallons in twenty-
four hours. Previously to this well being made, a trial boring,
of which Figs. 262, 263, are sections, was made. This boring
was lined with iron tubes 9 inches in diameter for 17 feet, inside
this 8 inches in diameter for 22 feet 9 inches, and within this
again a 5-inch tube. It was continued by a 5-inch bore reduced
to $4\frac{1}{2}$ inches, and at bottom to 3 inches.

Liverpool.—The oldest wells are at Bootle, to the north of the
town; these consisted in the first instance of three lodges or excavations in the rock, covering about 10,000 feet super. and about
$26\frac{1}{2}$ feet deep. These were covered with timber or slate roofs,
and in them sixteen bore-holes were sunk, of various diameters
and at depths ranging from 13 feet to 600 feet. In 1850 the
yield of one of these bore-holes was 921,192 gallons in twenty-
four hours, and the total yield in the same time only 1,102,065.
The water was collected in the lodges and conveyed by a tunnel
255 feet to a well 8 feet in diameter and 50 feet deep, from
which it was pumped. The yield of the Bootle well in 1865
was 643,678 gallons a day. Since this time a new well of oval
form, 12 feet by 9 feet and 108 feet deep, has been sunk, and at
its completion the yield rose to 1,575,000 gallons a day, but it
has again diminished considerably.

The Green Lane wells were commenced in 1845, the surface

being 144 feet above the sea level and their depth 185 feet, or 41 feet below the sea level. Headings extend in all about 300 feet from the shafts in various directions, three separate shafts being carried up to the surface. At first the yield was 1,250,000 gallons a day. A bore-hole, 6 inches in diameter, was then driven to a depth of 60 feet from the bottom of the well, when the yield increased to 2,317,000 gallons. In June 1856, the bore-hole was widened to 9 inches and carried down 101 feet farther, when the yield amounted to its present supply of over 3,000,000 gallons a day.

The large quantity of water yielded by the Green Lane wells is probably due to the existence of a large fault which is considered to pass in a north-westerly direction by the wells. In 1869 a bore-hole, 24 inches in diameter at the top and diminishing to 18 inches in diameter, was sunk from the bottom of a new shaft, 174 feet deep, to a depth of 310 feet, and the additional quantity of water derived from the new hole was about 800,000 gallons a day.

The Windsor Station well is of oval form, 12 feet by 10 feet and 210 feet deep, with a length of headings of 594 feet, and a bore-hole 4 inches in diameter and 245 feet deep. The yield is 980,000 gallons a day.

The Dudlow Lane well is also oval, 12 feet by 9 feet, and is sunk to a depth of 247 feet from the surface of the ground. Headings have been driven from the bottom of the well for a total distance of 213 feet, and an 18-inch bore-hole has been sunk to a depth of 196 feet from the bottom of the well, which is chiefly in a close hard rock, with occasional white beds from which the water is mainly obtained. The yield is nearly 1,500,000 gallons a day.

The total weekly supply from wells in Liverpool is upwards of 41,000,0 0 gallons, and there are also a great number of private wells drawing water from the sandstone, and their supply may be roughly estimated at 30,000,000 gallons a week.

Longton.—Staffordshire. The Potteries obtain a portion of their supply from a series of wells at Longton, which are

TRIAL BORING FOR WELL AT LEAMINGTON.

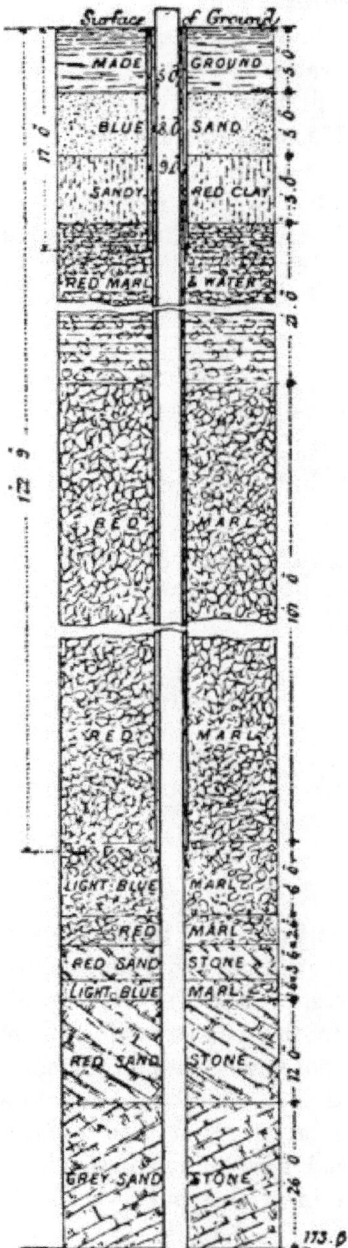

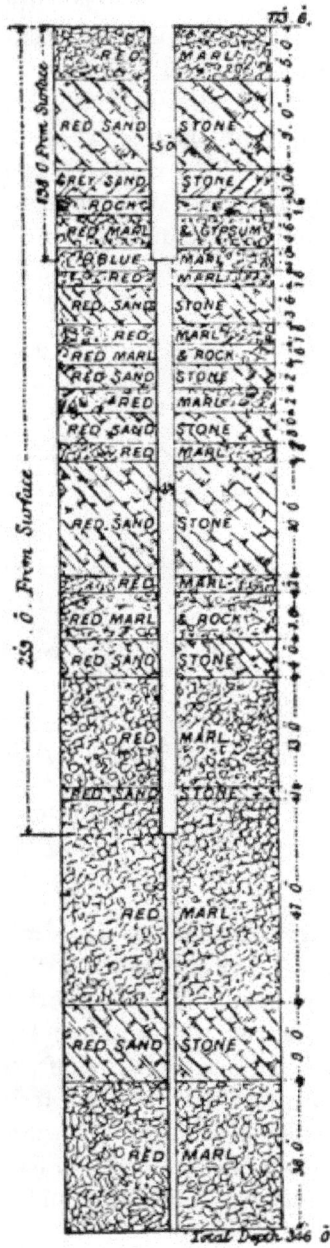

Fig. 262.

Fig. 263.

shown in the diagrammatic sectional plan, Fig. 264. The well marked No. 1 is 12 feet in diameter, and 135 feet deep in the new red sandstone. When finished, the water rose to within 35 feet from the surface. The cost of the first 45 feet was 3l. 10s. a yard ; of the second 45 feet, 6l. 10s. a yard ; and the third 45 feet, 9l. a yard. When this well was 36 feet down, a large quantity of water was met with, so a heading was driven at that depth in the direction of No. 2 well ; this, after 30 feet, passed

PLAN OF WELLS AT LONGTON.

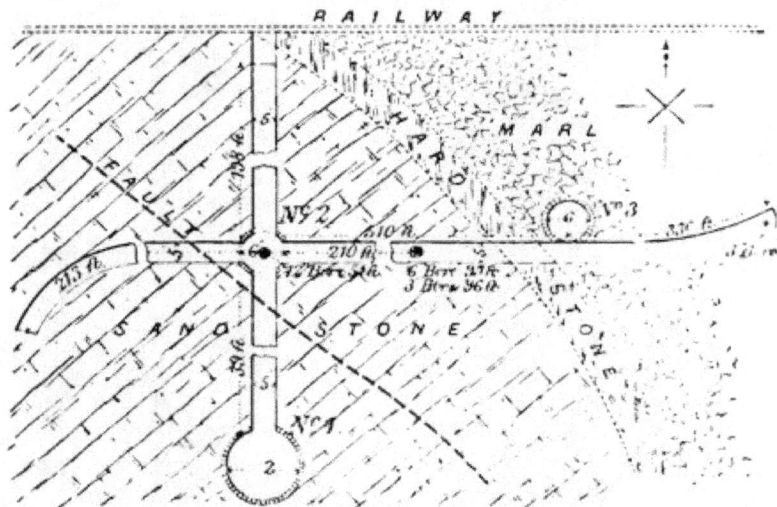

Fig. 264.

through a fault which drained off the water, and the sinking of No. 1 was proceeded with. After the engine had been erected and pumping some short time, it was proposed to drive headings from the bottom ; but owing to the pumps taking up so much room in the shaft, there was not space enough for sinking operations to be carried on, and No. 2 well was therefore sunk for convenience sake, at the cost of about 30s. a yard. When No. 2 was down 54 feet, a trial bore-hole 3 inches diameter was put down, and water rose in a jet about 3 feet high.

The well was then continued to the level of No. 1, and a heading, 39 feet long, driven between the two shafts. No. 2 has now a 12-inch bore-hole at bottom.

Headings have also been driven W. and N. of No. 2 well, at a cost of 30s. a yard. The western heading is 213 feet long, driven with a slight rise, and gave much water. There are two headings N., running in the direction of the railway, one over the other. The lower was driven level with the bottom of the shaft, but no water met with; the upper is 36 feet from the surface, and is intended to carry away surplus water down to a line of earthenware pipes which are led along the railway to a low-level reservoir.

In the eastern heading there is a rise of 4 feet owing to the nature of the strata; and after it had been driven 510 feet, well No. 3 was sunk for ventilation and for drawing out material. A bed of very hard sandstone, 63 feet long, was passed, cost 4l. 10s. a yard, and beyond came marl, in which driving cost 45s. a yard. This heading was continued 330 feet beyond No. 3, and an air-hole 3 inches diameter put down 126 yards deep, but no water was met with. The bed of hard sandstone was also found in driving the lower N. heading, which was discontinued after going into it some 5 or 6 feet. The yield from these wells is about 600,000 gallons a day, and recently a new bore-hole at No. 3 well, when down 350 feet, gave some 380,000 gallons a day additional.

Leek.—The Potteries' waterworks have also wells at the Wallgrange Springs, near Leek; these rise from the conglomerate beds, and are stated to yield 3,000,000 gallons daily. The water from these springs is pumped into Ladderidge reservoir, and is distributed from thence into the town of Newcastle-under-Lyme and the Potteries.

Middlesborough.—The Figs. 265 to 268 are sections and plans of a well at the works of Messrs. Bolckow and Vaughan, Middlesborough. A trial hole was first put down to a depth of 398 feet 6 inches, and a shaft afterwards sunk by Messrs. Docwra and Son to that depth, through alternating beds of clay, sand, gypsum, and sandstone. At the bottom of the shaft a bore-

WELL AT BOLCKOW AND VAUGHAN'S, MIDDLESBOROUGH.

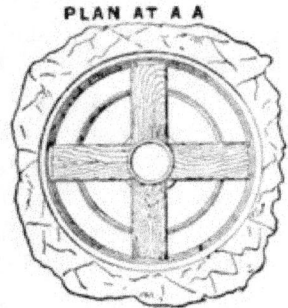

PLAN AT A A

Fig. 267.

Fig. 265.

hole of 18 inches diameter throughout was made with Mather and Platt's apparatus to a depth of 1312 feet ; the first 1160 feet of which were through new red sandstone interspersed with beds of clay, white sandstone, red marl, and gypsum. Next came 40 feet of gypsum, hard white sandstone, and limestone ; and the remaining 100 feet were through red sandstone, pure salt rock, occasional layers of limestone, and then salt rock to the bottom. The gross time spent in sinking this bore-hole was 510 days, or an average

PLAN AT B.B.

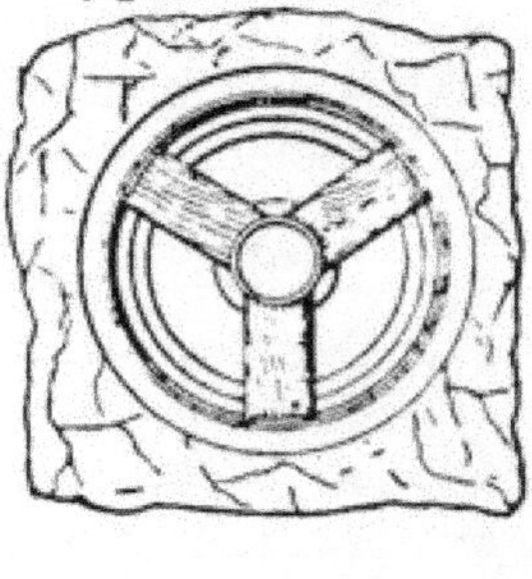

Fig. 268.

Well at Bolckow and Vaughan's, Middlesborough.

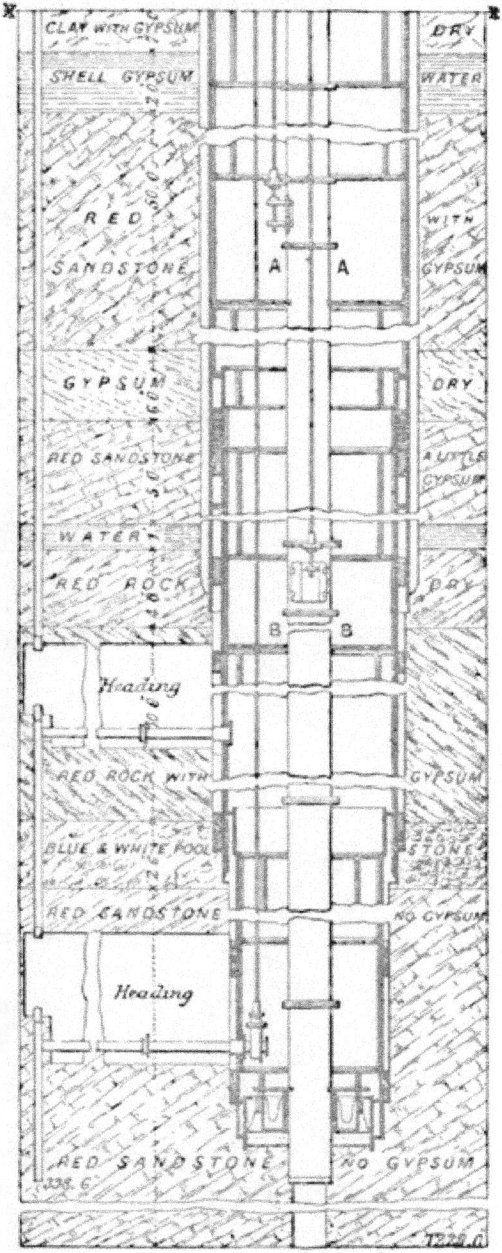

Fig. 266.

of 2 feet 5 inches a day.

Ross. — Herefordshire. The well at the Alton Court Brewery is shown in Figs. 269, 270. The shaft, 5 feet in diameter and 27 feet deep, is steined with 9-inch brickwork for a distance of 17 feet. At the bottom is a 12-inch bore-hole 100 feet 9 inches deep, unlined. The water is abundant. At level of the bore a heading, 6 feet high, 5 feet wide, and 27 long, has been driven, to afford storage room.

Wolverhampton.— This town is partially supplied from wells sunk in the

new red sandstone.   There are two shafts, 7 feet in diameter and 300 feet deep, a heading 459 feet long, and in this a boring of 390 feet.   The yield when first completed was 211,000 gallons a day.

Fig. 269.

WELL AT ROSS, HEREFORDSHIRE.

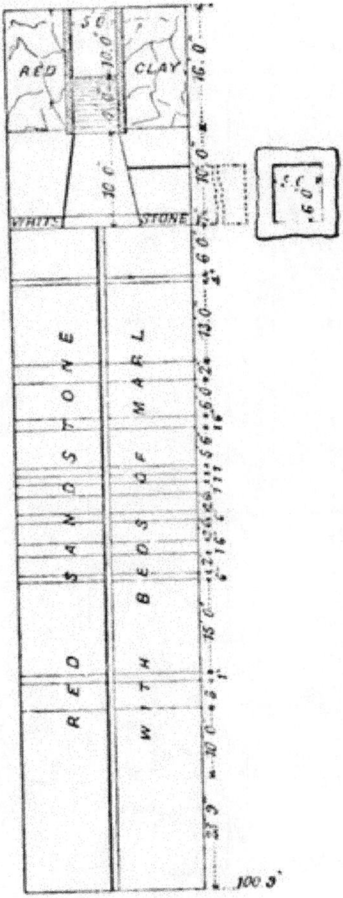

Scarborough. — The water-works well is at Osgodby; it is about 160 feet above the sea level.   The shaft is 10 feet diameter, and 91 feet deep, continued by a 6-inch bore-hole 136 feet deep.   There are three headings, of a total length of 70 yards.   The yield is from 600,000 to 800,000 gallons in the 24 hours; the water level varies, but is nor-mally 70 feet from surface. The surface-springs in the cover of drift have been en-tirely excluded by backing the steining of the shaft with puddle.

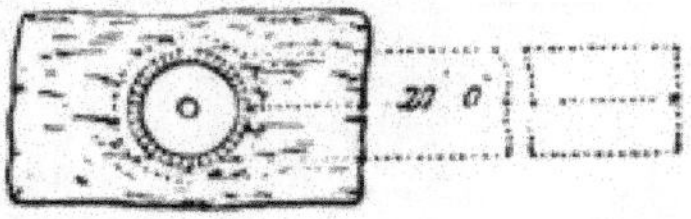

St. Helens. — Lancashire. Supplied with about 1,750,000 gallons of water daily from two wells, each about 210 feet deep; the well at Ecclestone Hill in pebble beds, and the well at Whiston in the lower mottled sandstone.   Each well has a bore-hole at the bottom.

OOLITIC STRATA.

Exeter.—Devonshire. Well at Silverton.   There is a shaft

Fig. 270.

5 feet diameter, and 20 feet deep, continued by a 6-inch bore-
hole to a total depth of 237 feet. Water level at 23 feet from
the surface; yield, 100 gallons a minute.

STRATA FROM BORE-HOLE:—	Feet.	In.
Sand	94	8
Rock	26	11
Marl	9	4
Clay and Greensand	30	0
Gravel	4	9
Hard Clay	16	0
Rock	15	10
Total	216	10

Northampton.—The well at the waterworks is sunk and bored
253 feet 3 inches in the lias. The shaft is steined with brickwork and iron cylinders in the following order: for 16 feet
9 inches in depth the well is 7 feet 6 inches in diameter, lined
with brickwork; at this depth two cast-iron cylinders 5 feet
6 inches diameter are introduced, which are again succeeded
by 9-inch brickwork, commencing at 5 feet 6 inches internal
diameter and widening out to 7 feet 6 inches in diameter.
The bottom of the shaft is floored with bricks at a distance of
120 feet from surface. At this point the bore-hole commences,
and for the first 31 feet it is lined with 14-inch pipes, which
rise into the shaft 5 feet above the floor. The remaining portion
of the bore-hole, 102 feet, is 9 inches diameter.

Selby.—Yorkshire. Well at waterworks consisting of a 6-inch
bore-hole 330 feet deep, yielding about 243,000 gallons in the
twenty-four hours, water level 4 feet from surface. The strata
passed were—

	Feet.	In.
Warp and Clay	10	0
Strong Clay	10	6
Sand and Clay	14	8
Strong Clay	7	10
Clay and Silt	8	9
Grey, or Loose Water Sand	7	9
Red Sand	6	6
Carried forward	66	0

	Feet.	In.
Brought forward	66	0
Indurated Sand	1	6
Red Sandstone	54	6
Red Clay and Fullers' Earth, with pipe-clay ..	5	0
Red Sandstone	203	0
Total	330	0

The pebble beds have been bored through at various points between Nottingham, Retford and Selby, and are directly overlain by the water-stones, the upper mottled sandstone and Keuper conglomerate being alike absent. North of Selby the pebble beds also have thinned out, and the Keuper water-stones and the lower mottled sandstone are alone available for underground water-supply in the plains of York.

York.—Well at Towthorpe Common, 60 feet above Ordnance datum. Consists of a 9-inch bore-hole 311 feet deep, this was subsequently plugged and reduced to 210 feet. Yield abundant. The section is as follows :—

	Feet.	In.
Top Sand	4	6
Fine Clay	15	0
Boulder Clay	15	0
Loamy Sand	6	0
Fine Warp Clay	9	0
Grey Sand	10	0
Boulder Clay	4	0
Greensand	16	0
Greensand, with layers of blue bind	18	0
Blue Bind or Marl	1	9
Light Green-and, with blue bind	35	0
White Sandstone	5	0
Blue Bind	1	0
Red Marl	2	0
White Sandstone	81	0
Blue Marl	0	6
White Sandstone	23	0
Blue Marl	0	3
Variegated Sandstone	60	0
Red Marl	3	0
Total	310	0

Salton, near Malton.—Well 150 feet above Ordnance datum, is a bore-hole 4 inches diameter, 316 feet deep; the water

flows out at surface. It passes made earth and about 15 feet
of fluviatile drift, continued by 295 feet of Kimmeridge clay.

Swanage.—Dorset. The section and
plan, Figs. 271, 272, are of a well at
Swanage, sunk 60 feet and bored 53
feet, the lining tube rising 8 feet into
the shaft, which is 5 feet 6 inches in
diameter, and lined with 9-inch steining. The strata passed through are
clays and limestones, and may perhaps
be referred to the Purbeck beds. At
first this well yielded little or no water,
but it now gives a sufficient supply.

Fig. 271.

WELL AT SWANAGE, DORSET.

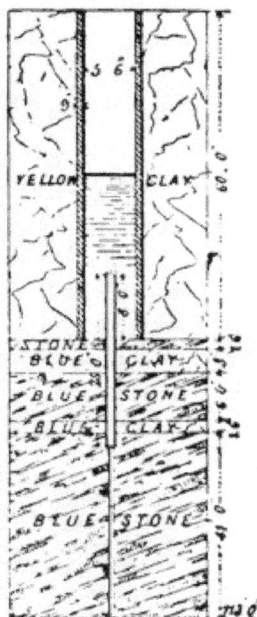

PLAN.

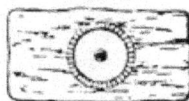

Fig 272.

CRETACEOUS STRATA.

Beccles.—Norfolk. Waterworks; the
wells are situated about three-quarters
of a mile S. of the town, at 100 feet
above sea level. Water first occurred
in the beds at 80 feet from surface of
good quality, at a farther depth of 10
feet the supply gave 21,000 gallons a
day without lowering the top water
level. This is the water-bearing stratum
generally around Beccles, most of the
wells being sunk into it. There are two
wells at the waterworks very similar in
section, both are carried into the chalk,
which yields an abundant water supply. The details given
below are of well No. 2, consisting of a shaft to a depth
of 91 feet continued by a 9-inch bore-hole ;—

STRATA ;—

	Feet.	In.
Vegetable Soil	1	0
Chalky Boulder Clay	10	3
Middle Glacial Beds	17	9
Carried forward	29	0

P

	Feet.	In.
Brought forward	29	0
Bure Valley Beds, Gravels, and Sands	33	0
Sands and Loam, with much iron	15	0
White Yellow Sands, with loam	14	6
Fluvio-Marine Crag	65	6
Chalk, with flints	73	0
Total	230	0

Beccles.—Well at Worthington and Co.'s ;—

	Feet.
Gravels and Sands	58
Chalk, upper part like pipe-clay	26
Total	84

Bishop Stortford.—The waterworks and well are situate W. of the town, near the farm buildings known as Marsh Barns. The shaft is 160 feet deep, the bore-hole 140 feet. The following is a section of the strata ;—

	Feet.
BOULDER CLAY	17
LONDON CLAY, 54 feet ;—	
Brown Clay	14
Black Clay	2
Black Sandy Loam, with iron pyrites	12
Black Clay, with lignite	11
Dark Grey Sand, with large pieces of sandstone and shells	15
READING BEDS, 45½ feet ;—	
Black Clay	2
Brown Clay	20
Light Brown Sand	1
Variegated Sand	18
Brown Clay	4
Flints and Pebbles	1
To Chalk	117
CHALK	183
Total	300

The water rises to within 140 feet of the surface of the ground. The yield is 10,000 gallons a minute; only 25 gallons a minute from the bore, the rest from the headings driven north and south respectively at a depth of 154 feet.

Braintree.—Essex. The well sunk for the Local Board is in a field near Pod's Brook. The shaft is 8 feet in diameter, steined with 9-inch steining, and carried down 55 feet, the remainder of the well being bored. Strata ;—

	Feet.
DRIFT, 14 feet :—	
Sandy Gravel	5
Drift Clay	9
LONDON CLAY, 136 feet :—	
Clay, with sand, shells, and septaria, the bottom part more sandy	126
Dark Sand, with a few shells, yielding much water	10
READING BEDS, 45 feet :—	
Mottled Plastic Clays, getting more sandy lower down, and with specks of chalk	44
Coarse Black Sandy Clay	1
THANET SAND (?), 33 feet :—	
Light-coloured Sands, firm and hard, getting darker and more friable lower down	20
Light-coloured Sands, firm, changing to coarse and dark	13
To Chalk	228
CHALK, with much water, rising to about 12 feet from the surface	17
Total	245

The level of the ground is 140 feet above the sea level ; water stands 29 feet deep ; yield about 11,500 gallons an hour.

Brighton.—This town has always been supplied from wells sunk in the chalk. One well is sunk near the Lewes Road, and has a total length of 2400 feet of headings driven in a direction parallel with the sea, and at about the coast level of low water. These headings intercept many fissures and materially add to the yield.

A second well was sunk in 1865 at Goldstone Bottom, and headings driven to the extent of about a quarter of a mile across the valley parallel to the sea.

Goldstone Bottom is a naturally formed basin in the chalk, the lowest side of which, nearest the sea, is more than 60 feet higher than the middle or bottom of the basin. The water is obtained as at Lewes Road, from fissures running generally

at right angles to the coast line, but they are of much larger
size and at far greater distances from each other; whereas at
the Lewes Road well it is rare that 30 feet of headings were
driven without finding a fissure, and the yield of the largest
was not more than 100 to 150 gallons a minute. At Goldstone
nearly 160 feet were traversed without any result, and then an
enormous fissure was pierced which yielded at once nearly 1000
gallons a minute; and the same interval was found between this
and the next fissure, which was of a capacity nearly as large. The
total length of the headings at Goldstone Bottom is 13,000 feet.
The yield from each well is about 3,000,000 gallons daily.

Bletchingly.—Surrey. Well at Highfield, and bore-holes in
the lower greensand, sunk under the writer's superintendence.

Water is found at 45 feet from surface at the house, and from
55 to 59 feet from surface in various parts of the grounds.
The yield abundant, so far as tested, upwards of 300 gallons an
hour.

SECTION AT BORE-HOLE No. 1.

	Feet.	In.
SURFACE SOIL	1	0
CLAY	5	0
SANDGATE BEDS, 65 feet :—		
Hard and Soft Sandstone	15	4
Brown Clay	5	4
Sandstone and Sand	14	3
Fullers' Earth, mixed with sand	2	8
Clay and traces of Fullers' earth	5	8
Dry and White Clay	1	0
Fullers' Earth	0	2
Blue and Grey Sandstone (*hard*)	5	3
Fullers' Earth	3	2
Clay and Sand	2	10
Fullers' Earth	4	6
Sandstone	2	10
Fullers' Earth	2	0
Total	71	0

Chelmsford.—The well belonging to the Local Board of
Health, situated at Moulsham, yields about 95,000 gallons of
water a day. It is sunk for 200 feet; the rest bored. Water
overflowed at first, but now that the well is in use and pumped

from, the water only rises to 76 feet from the surface. The
following strata were pierced ;—

	Feet.	In.
BLACK SOIL (Mould)	3	0
DRIFT, 63½ feet:—		
Yellow Clay..	2	6
Gravel	12	6
Quicksand	44	6
Sand with Stones	4	0
LONDON CLAY, 186½ feet:—		
Clay	104	0
Clay, with sand	50	0
Dark Sand	12	6
Clay Slate (? septaria)	0	9
Clay and Shells	4	0
Clay Slate (? septaria)	0	3
Dark Sand and Clay	9	6
Sand and Shells	4	0
Pebbles	1	6
WOOLWICH BEDS;—		
Sand	7	0
Red Clay	12	0
Clay and Sand	64	0
DARK THANET SAND	30	0
To Chalk	366	0
CHALK, 202 feet;—		
Chalk	88	0
Rubble	1	0
Chalk	113	0
Total	568	0

Cheshunt, New River Company.—Situate at the engine-house
between the two reservoirs. The well is 171 feet deep, and is
steined partly with brickwork and partly with iron cylinders.
For 12 feet in depth the well is 11 feet 6 inches in diameter,
and steined with 14-inch brickwork; for a farther depth of a
few feet it is 9 feet diameter, and steined with 9-inch brickwork; it is then lined with cast-iron cylinders, 8 feet diameter,
which are carried to a depth of 105 feet from the surface.
There are fifteen cylinders of this size in use, and they are succeeded by others 6 feet 10 inches diameter, of which there are
six in use; these are again succeeded by two cylinders 6 feet
diameter. The whole of the cylinders are 6 feet in depth. The

bottom of the last cylinder is 118 feet from the surface, at which point they rest upon a foundation of 9-inch brick steining 7 feet in depth. At the bottom of the 6-feet cylinders the well widens out in the form of a cone 12 feet 6 inches diameter at the floor, which is 26 feet below the bottom of the 6-feet cylinder. In the centre of the well a bore-hole, 3 inches diameter and 27 feet deep, was made, and the well is provided on the floor level with headings.

SECTION OF STRATA.

	Feet.	In.
SURFACE EARTH	1	6
GRAVEL..	8	0
LONDON CLAY, 47 feet ;—		
Blue Clay	45	0
Yellow Clay	2	0
READING BEDS, 51 feet ;—		
White Sand	12	0
Dark Sand	39	0
To Chalk	107	6
CHALK	63	6
Total	171	0

Dorking, Surrey, obtains its water supply from a well sunk into the outcrop of the lower greensand, at the south side of the town. The shaft is 11 feet in diameter and 160 feet deep, steined with 9-inch work laid dry. The yield is not more than 30 gallons a minute, owing to the unfortunate position of the well, but might be considerably increased if suitable means were adopted.

Harrow Waterworks.—The well is situate 430 yards to the west of the church. The surface of the ground is 226 feet above the Ordnance datum. There is a shaft for $193\frac{1}{2}$ feet; the rest is a bore. In a bed of dark red sand 144 feet down, the water was very foul. Strata ;—

	Feet.	In.
Light Blue Clay, with light-coloured stone	19	11
Brown Clay, with white stone	54	11
Dark Mottled Clay	15	0
Similar Clay, with dark and green sand	4	0
Carried forward	93	10

	Feet.	In.
Brought forward	93	10
Dark mottled clay, very hard..	3	0
The same, very hard, and dark sand ..	2	0
Lighter-coloured Hard Clay	5	0
The same, and dark sand	6	6
Large Pebbles	0	6
Clay and Sand	5	0
Light Blue Clay	0	4
Light-coloured Stone, with red and blue spots	1	3
Mottled Clay	7	11
Yellow, Light Blue, and Green Clay ..	1	0
Dark Green Clay, with black veins and spots	5	0
Blue Clay	1	6
Very Hard Brown, Yellow, and Blue Clay	4	0
Light Brown Running Sand, with water	2	6
Hard Mottled Clays..	6	6
Light Brown Dead Sand..	8	8
Black Peat, with dark pebbles	0	6
Brown and Green Gravel, with flints ..	3	2
Green Clay	0	4
To Chalk	158	6
Chalk, with beds of flint 4 to 15 inches in thickness, 15 to 24 inches apart; 395½ feet down, from surface, a bed of flint 6 feet thick	254	0
Total	412	6

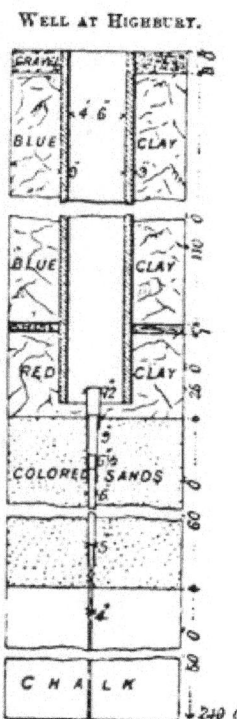

WELL AT HIGHBURY.

Fig. 273.

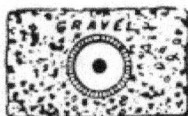

Fig. 274.

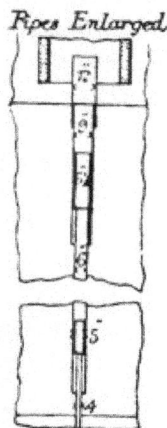

Pipes Enlarged

Fig. 275.

Water rises to a height of 125 feet below the surface. The yield is about 190 gallons a minute.

Highbury.—Middlesex. Well at the residence of H. Rydon, Esq., New Park, Figs. 273 to 275. The shaft is 4 feet 6 inches diameter, and 136 feet deep, steined with 9-inch work set in cement. The bore was commenced with a 12-inch hole, but the character of the ground was such that the successive reductions in size, shown in the enlarged section of the lining tubes, Fig. 275, had to be

made.   When in the chalk the bore was continued some 48 feet
unlined.   The strata passed were ;—

		Feet.
GRAVEL		3
LONDON CLAY, 111 feet ;—		
Blue Clay		110
Claystone		1
READING AND THANET SAND, 85 feet ;—		
Mottled Clay		25
Coloured Sand		60
To Chalk		199
CHALK		50
Total		249

Holloway.—Middlesex.   Well at Islington Workhouse, shaft
234 feet, continued by a bore-hole, gives 25,000 gallons daily ;—

STRATA ;—		Feet.
London Clay		230
Woolwich and Reading Beds		51
Pebbles		1
Thanet Sand		16
Green Flints		2
Chalk		248
Total		548

Kentish Town.—This well was sunk under the supposition
that as the outcrop of the subcretaceous formations was continuous around the margin of the cretaceous basin surrounding
and underlying the London tertiaries, except at the eastern
border, those subcretaceous formations would be found under
London, just as they actually were at Paris.   This proved to
be the case until the gault was passed, when a series of sandstones and clays were encountered, occupying the place of the
lower greensand, but evidently of older geological character
and having many of the features of the new red sandstone.
Confirmation of these views has been since given by the clear
section obtained at Meux' Brewery, Tottenham Court Road,
shown on p. 234.

The surface of the ground, Fig. 276, is 174 feet above
Thames high-water mark.   There is a shaft for 539 feet; the

BORING AT KENTISH TOWN, LONDON.

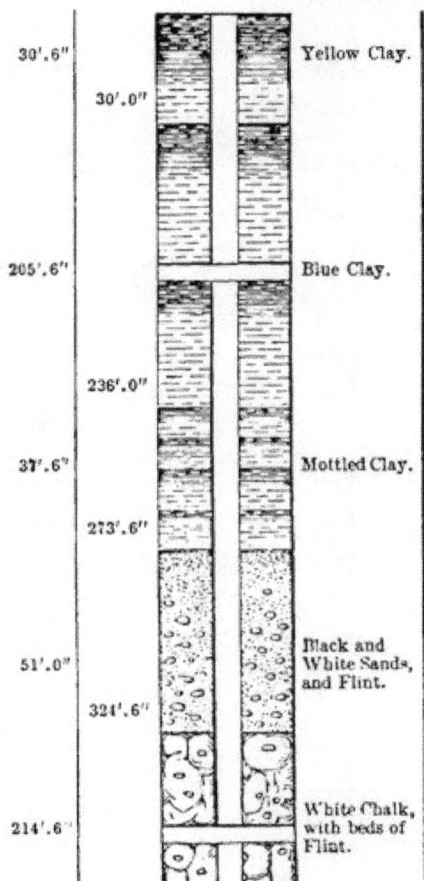

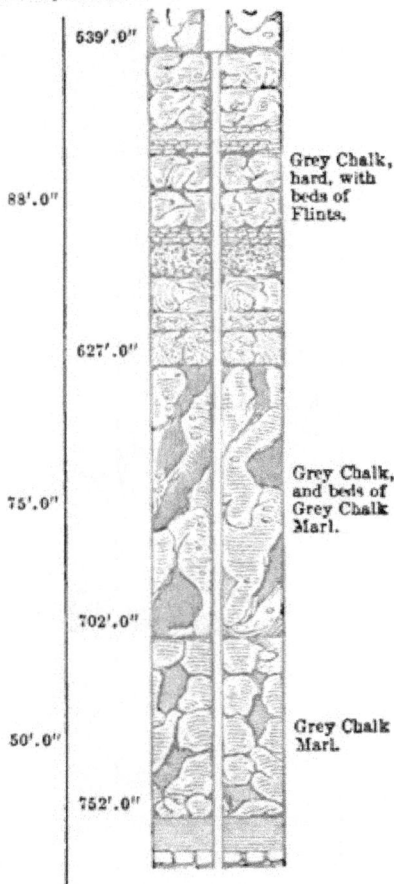

Fig. 276.

Fig. 277.

remainder being bored. The following detailed account of the
strata is due to Prestwich ;—

		Feet.	In.
LONDON CLAY, 236 feet ;—			
Yellow Clay		30	6
Blue Clay, with septaria		205	6
READING BEDS, 61½ feet ;—			
Red, Yellow, and Blue Mottled Clay		37	6
White Sand, with flint pebbles		0	6
Carried forward		274	0

BORING AT KENTISH TOWN, LONDON—*continued.*

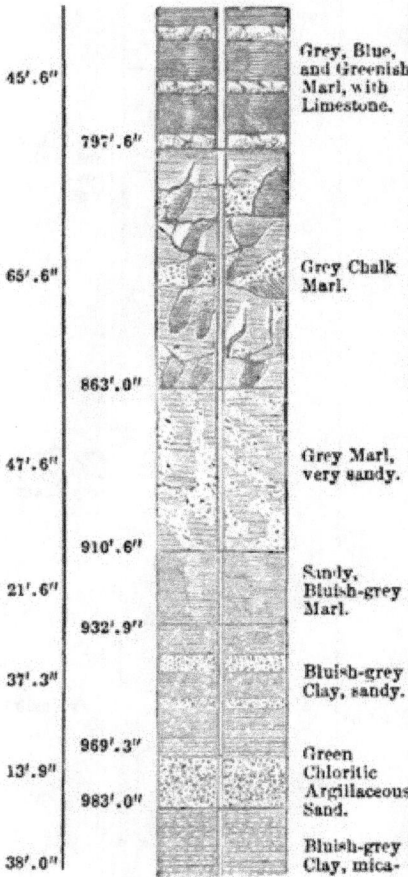

Fig. 278.

45'.6"

797'.6"

Grey, Blue, and Greenish Marl, with Limestone.

65'.6"

863'.0"

Grey Chalk Marl.

47'.6"

910'.6"

Grey Marl, very sandy.

21'.6"
932'.9"

Sandy, Bluish-grey Marl.

37'.3"

969'.3"

Bluish-grey Clay, sandy.

13'.9"
983'.0"

Green Chloritic Argillaceous Sand.

38'.0"

Bluish-grey Clay, mica-

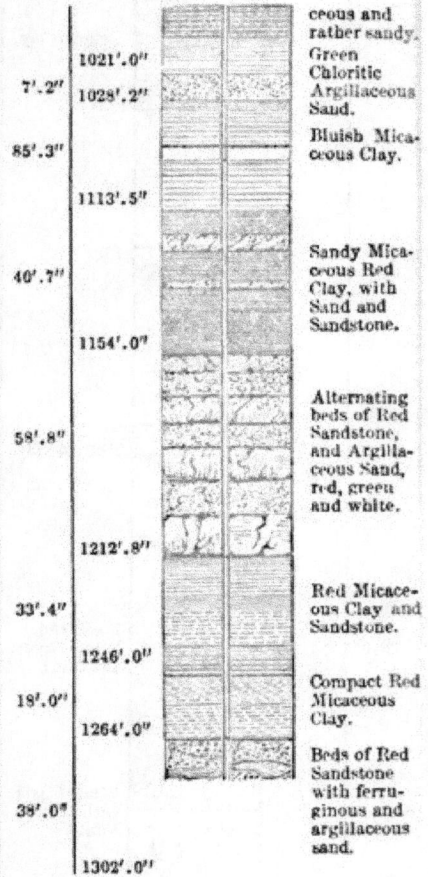

Fig. 279.

1021'.0"
7'.2"
1028'.2"

85'.3"

1113'.5"

40'.7"

1154'.0"

58'.8"

1212'.8"

33'.4"

1246'.0"
18'.0"
1264'.0"

38'.0"

1302'.0"

ceous and rather sandy. Green Chloritic Argillaceous Sand.

Bluish Micaceous Clay.

Sandy Micaceous Red Clay, with Sand and Sandstone.

Alternating beds of Red Sandstone, and Argillaceous Sand, red, green and white.

Red Micaceous Clay and Sandstone.

Compact Red Micaceous Clay.

Beds of Red Sandstone with ferruginous and argillaceous sand.

	Feet.	In.
Brought Forward	274	0
READING BEDS, *continued*;—		
Black Sand, passing into the bed below	2	0
Mottled Green and Red Clay	1	0
Clayey Sand	3	0
Dark Grey Sand, with layers of clay	9	6
Ash-coloured Quicksand	6	6
Flint Pebbles	1	6
Carried forward	297	6

	Feet.	In.
Brought forward	297	6

THANET SAND, 27 feet ;—

Ash-coloured Sand	10	0
Clayey Sand..	4	0
Dark Grey Clayey Sand	11	0
Angular Green-coated Flints	2	0

CHALK, WITH FLINTS (? UPPER CHALK), 244½ feet ;—

Chalk, with flints	119	6
Hard Chalk, without flints	8	0
Chalk, softer, with a few flints..	31	6
Nodular Chalk, with three beds of tabular flints ..	13	6
Chalk, with layers of flint	32	6
Chalk, with a few flints and patches of sand	9	6
Very Light-grey Chalk, with a few flints	30	0

CHALK, WITHOUT FLINTS (LOWER CHALK), 341 feet ;—

Light Grey Chalk, and a few thin beds of marl ..	133	0
Grey Chalk Marl, with compact and marly beds and occasional pyrites	161	0
Grey Marl	20	0
Harder Grey Marl, rather sandy and with occasional pyrites	27	0

CHALK MARL, 59¼ feet ;—

Hard Rocky Marl (? Tottenhoe Stone)	0	6
Bluish Grey Marl,rather sandy,lower part more clayey	58	9

UPPER GREENSAND :—

Dark Green Sand, mixed with grey clay	13	9

GAULT, 130½ feet ;—

Bluish Grey Micaceous Clay, slightly sandy	39	0
The same, with two layers of clayey greensand ..	6	7
Micaceous Blue Clay ; at base a layer full of phosphatic nodules	84	11

LOWER GREENSAND (?), 188½ feet ;—

Red and Yellow Clayey Sand and Sandstone ..	1	0
Compact Red Clay, with patches of variegated sandstone	4	0
Dark Red Clay	4	7
Red Clay, Whitish Sand, and Mottled Sandstone..	3	0
Hard Red Conglomerate, with pebbles from the size of a marble to that of a cannon-ball	2	0
Micaceous Red Clay, mottled in places	26	0
Layers of White Sandstone and Red Sand	3	8
Mottled Sandstone	0	4
Red Sand and Sandstone, with pebbles (a spring)	2	0
Layers of Red Sandstone and White Sand	4	0
Pebbly Red Sand and Sandstone	1	0
White and Red Sandstone	5	0
Fine Light Red Sand..	2	9
Hard Sandstone	0	3

Carried forward..	1173	1

	Feet.	In.
Brought forward	1173	1

LOWER GREENSAND (?), *continued ;—*

	Feet.	In.
Very Fine Light Red Sand	4	0
Red Clay	2	0
Clayey Sand	1	3
Red Sandy Micaceous Clay, with sandstone	2	5
Compact Hard Greenish Sandstone	10	0
Very Micaceous Red Clay	1	0
Grey and Red Clayey Sand	1	1
Light-coloured Soft Sandstone	2	1
Red Sand and Sandstone	6	2
Greenish Sandstone	4	0
White and Grey Clayey Sand, with iron pyrites ..	2	0
Reddish Clayey Sand, with layers of sandstone ..	3	8
Micaceous Red Clay	18	4
Greenish Sandstone	0	5
Red Mottled Micaceous Clay, with patches of sand	34	6
Red Quartzose Micaceous Sandstone	2	0
Brownish-red Clayey Sand and Sandstone	4	0
Very Hard Micaceous Sandstone, with pebbles of white quartz	4	0
Light Red Clayey Sand	10	0
Red Micaceous Quartzose Sandstone	8	0
Light Red Clayey Sand, small fragments of chalk	2	0
Whitish and Greenish Hard Micaceous Sandstone	6	0
Total..	1302	0

The engravings, Figs. 276 to 279, which are on the authority of G. R. Burnell, do not exactly agree with Prestwich's section, but in the main they are both alike. The following summary may be found of service ;—

	Feet.	In.
London Clay	236	0
Lower London Tertiaries	88	6
Chalk	644	9
Upper Greensand	13	9
Gault	130	6
Lower Greensand (?)	188	6

Limehouse.—Middlesex. Well at Taylor and Walker's Brewery. Consists of a shaft for 143 feet lined with 7 feet 6 inches and 5 feet cylinders, and continued by a 12-inch bore-hole 157 feet deep.

SECTION OF STRATA.

	Feet.
Made Earth and Loam	14
VALLEY DRIFT ;—	
Gravel and Sand	15
Carried forward	29

	Feet.
Brought forward	29
LONDON CLAY, 44 feet ;—	
Sandy Blue Clay	2
Brown Clay	19
Sand	8
Blue Clay with Shells	11
Blue and Green Sand	4
WOOLWICH BEDS, 10 feet ;—	
Pebbles	6
Sand	4
THANET SANDS, 58 feet ;—	
Green Sand and White Pebbles	14
Grey Sand	44
To Chalk	141
Chalk Flints	2
Hard and Soft Chalk	153
Total	296

Loughton.—Essex . Bore-hole at Great Eastern Railway Station.

STRATA :—	Feet.	In.
Tertiaries	243	0
Chalk	648	6
Chalk Marl and Upper Green Sand	37	0
Gault	132	6
„ with Pebbles	31	6
Total	1092	6

Michelmersh.—Hants. Fig. 280 shows a section of a well in this village, comprised within the writer's practice. The shaft is 4 feet 6 inches in diameter and 400 feet deep, steined both above and below the chalk with 9-inch work, the upper course having rings of cement at every 12 inches.

The strata pierced were ;—

	Feet.	In.
Surface Soil	4	0
Dark Clay	27	0
Chalk	250	0
Band of Calcareous Sand	2	6
Upper Greensand	17	0
Total	300	6

The water rises some 19 feet in the shaft, and is abundant, although up to the present its quantity has not been tested.

Mile End.—Middlesex.   Well at Charrington, Head, and Co.'s brewery, Figs. 281 to 283.   The surface is 33½ feet above Trinity high-water mark.

In the upper part there are three iron cylinders built upon 9-inch brickwork, which is carried down into the mottled clay. A 9-inch iron cylinder, partially supported by rods from the surface, rises some 28 feet into the brick shaft into which it is built by means of rings.   Another iron cylinder is carried down into the chalk, the space between the cylinders being filled in with concrete.

The strata passed were ;—

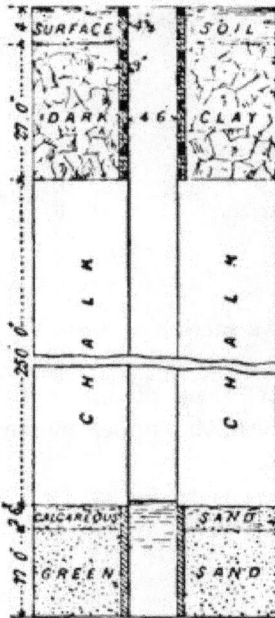

WELL AT MICHELMERSH.

Fig. 290.

	Feet.	In.
MADE EARTH	7	0
VALLEY DRIFT, 6 feet ;—		
Sand	3	0
Gravel	3	0
LONDON CLAY, 86 feet ;—		
Blue Clay	7	0
Hard Brown Clay, with claystones	68	0
Brown Sandy Clay	2	0
Hard Brown Sandy Clay, rotten at bottom	9	0
WOOLWICH AND READING BEDS	63	0
THANET SAND, 40 feet ;—		
Green Sand	2	0
Brownish-green Quicksand and Pebbles	2	0
Brown Sand	2	0
Grey and Brownish-green Sand	2	0
Green Sand and Pebbles ..	2	0
Brown Sand	2	0
Green Sand and Pebbles ..	15	0
Grey Sand and small Pebbles	2	0
Dark Grey and Green Sand	10	6
Green Sand and Green-coated Flints	0	6
To Chalk	202	0
Chalk Flints	0	6
Hard Chalk and Water ..	2	0
Total	204	6

WELL AT CHARRINGTON'S, MILE END.

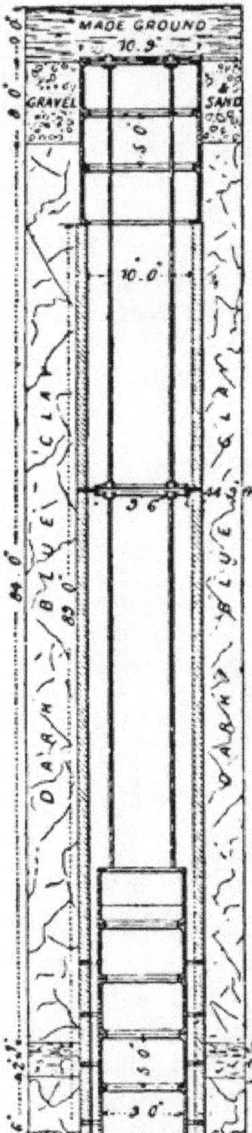

Fig. 281.

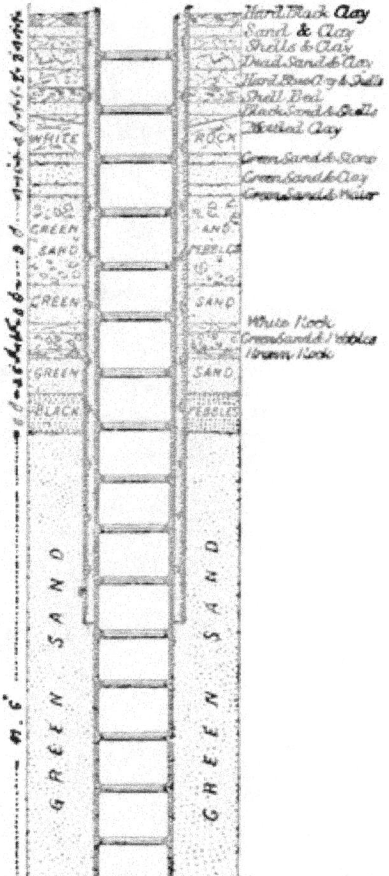

Fig. 282.

PLAN.

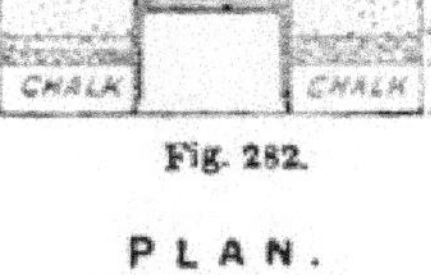

Fig. 283.

The water level is some 103 feet from surface, and the yield 60,000 to 70,000 gallons a day.

Norwich.—Well at Coleman's works. After a few feet of alluvium, the borer passed through hard chalk with flints at distances of about 6 or 7 feet apart, for 700 feet, with the exception of 10 feet at the depth of 500 feet where the rock was soft and of a rusty colour, thence the flints were thicker, namely, about 4 feet apart to the depth of 1050 feet. After this 102 feet were pierced of chalk, free from flints, to the upper greensand, a stratum of about 6 feet, and then gault for 36 feet. The whole boring being full of water to within 16 feet of the surface.

Section of strata :—	Feet.
Alluvium	12
Hard Chalk, with flints	483
Soft Chalk	10
Hard Chalk	190
Hard Chalk, flints closer	350
Chalk without flints	102
Upper Greensand	6
Gault	36
Total	1189

Norwich, various wells at. The water in most cases is abundant.

	Tertiary Strata. Feet.	Chalk. Feet.	Total. Feet.
Distillery	48	228	270
Morgan's Brewery	12	218	230
Pockthorpe ,,	20	230	250
Rosary Cemetery	50	50	100
Mousehold, at farm	90	42	132
,, ,, J. Harvey's	60	100	160

Paris.—The wells sunk in the Paris basin, of which Fig. 284 is a section, are very numerous, and many of them of great depth. Fig. 285 is a plan indicating the position of the principal wells, and Figs. 286 to 288 sections giving each a summary of the nature and thickness of the formations passed through.

For boring these wells special tools had to be used, which have already been described at length in Chap. VII.

A large Artesian well, constructed by Dru at Butte-aux-Cailles, for the supply of the city of Paris, was intended to be carried down through the greensand to a depth of 2600 or 2900 feet to reach the Portland limestone. The boring was suspended in 1872 for municipal reasons, it was then 1745 feet deep, and its diameter 47¼ inches. Its section for the first 496 feet is shown in Fig. 284.

During the previous years, M. Dru was engaged in sinking a similar well of 19 inches diameter for supplying the Sugar Refinery of M. Say, in Paris, Fig. 285; 1570 feet of this well had been bored in 1867, see Fig. 288. It was finished in 1869, at a total depth of 1903 feet, the bottom of the bore-hole being in the lower greensand. It yields 1760 gallons a minute.

The well at Grenelle was sunk by Mulot in 1832, and after more than eight

GEOLOGICAL SECTION FROM NIORT TO VERDUN, THROUGH THE PARIS BASIN.

Horizontal scale, 90 miles the inch.
Vertical scale, 1500 feet the inch.

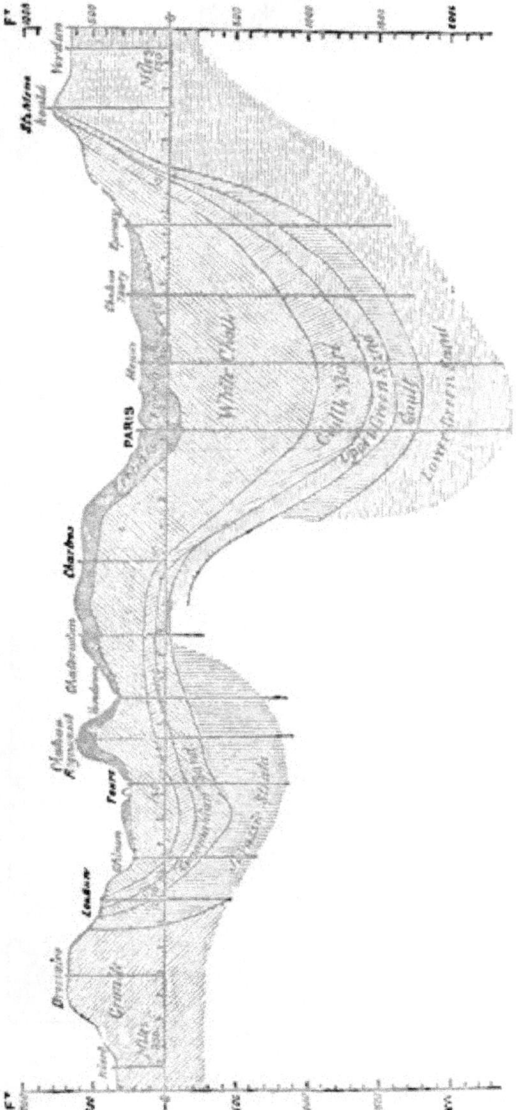

Fig. 284.

Q

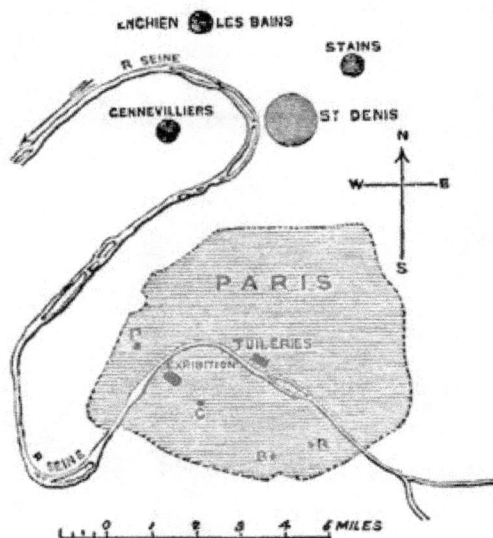

Fig. 285.

References.—P. Passy. G. Grenelle. B. Butte-aux-Cailles.
R. Sugar Refinery.

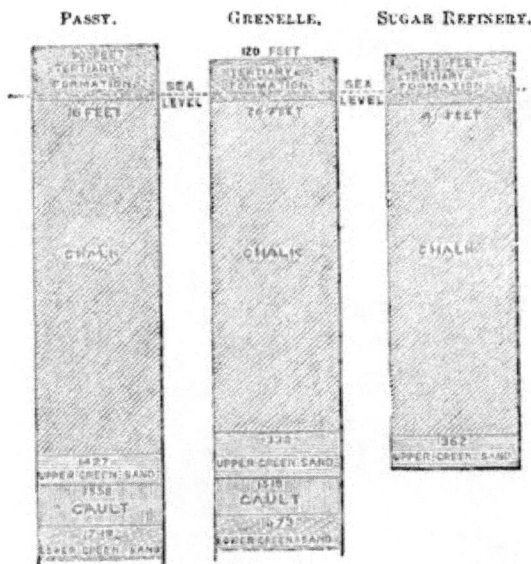

Fig. 286.

Fig. 287.

Fig. 288.

years' incessant labour, water rose on the 26th of February, 1842, from the total depth of 1806 feet 9 inches. The diameter of the borehole is 8 inches, ending, as is seen in the detail sections, Figs. 289 to 292, in the lower greensand.

The well of Passy was intended to be executed in the Paris basin which it was to traverse with a diameter, hitherto unattempted, of 1 mètre (3·2809 feet); that of the Grenelle well being only 20 centimètres (8 inches). It was calculated that it would reach the water-bearing stratum at nearly the same depth as the latter, and would yield 8000 mètres or 10,000 cubic mètres in twenty-four hours, or about 1,786,210 to 2,232,800 gallons a day.

Figs. 293 to 296 show a detail section of the strata passed.

The operations were undertaken by Kind under a contract with the Municipality of Paris, by which he bound himself to complete the works within the space of twelve months from the date of their commencement, and to deliver the above

BORING AT GRENELLE, PARIS.

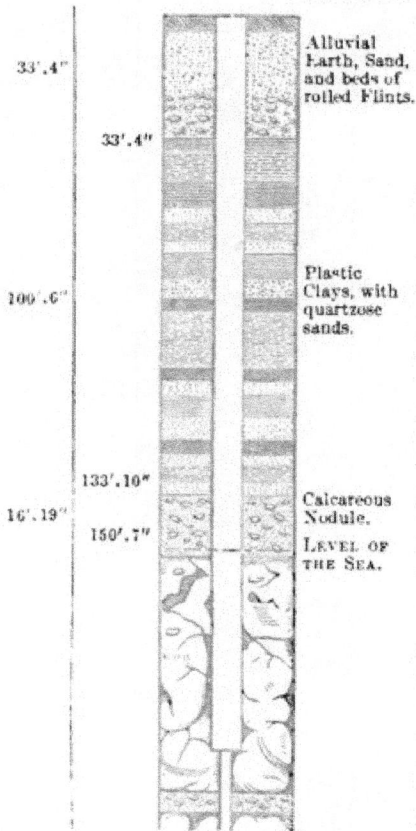
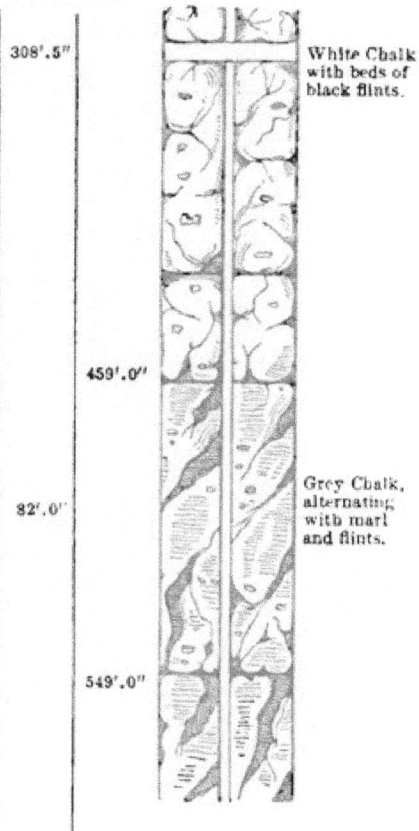

33′.4″

33′.4″

100′.6″

133′.10″

16′.19″

150′.7″

Alluvial Earth, Sand, and beds of rolled Flints.

Plastic Clays, with quartzose sands.

Calcareous Nodule.

LEVEL OF THE SEA.

308′.5″

459′.0″

82′.0″

549′.0″

White Chalk with beds of black flints.

Grey Chalk, alternating with marl and flints.

Fig. 289.   Fig. 290.

quantity of water for the sum of 300,000 francs, 12,000l. On the 31st of May, 1857, after the workmen had been engaged nearly the time stipulated for the completion of the work, and when the boring had been advanced to the depth of 1732 feet from the surface—the excavation suddenly collapsed in the

upper strata, at about 100 feet from the ground, and filled up
the bore. Kind would have been ruined had the engineers of
the town held him to the strict letter of his contract ; but it

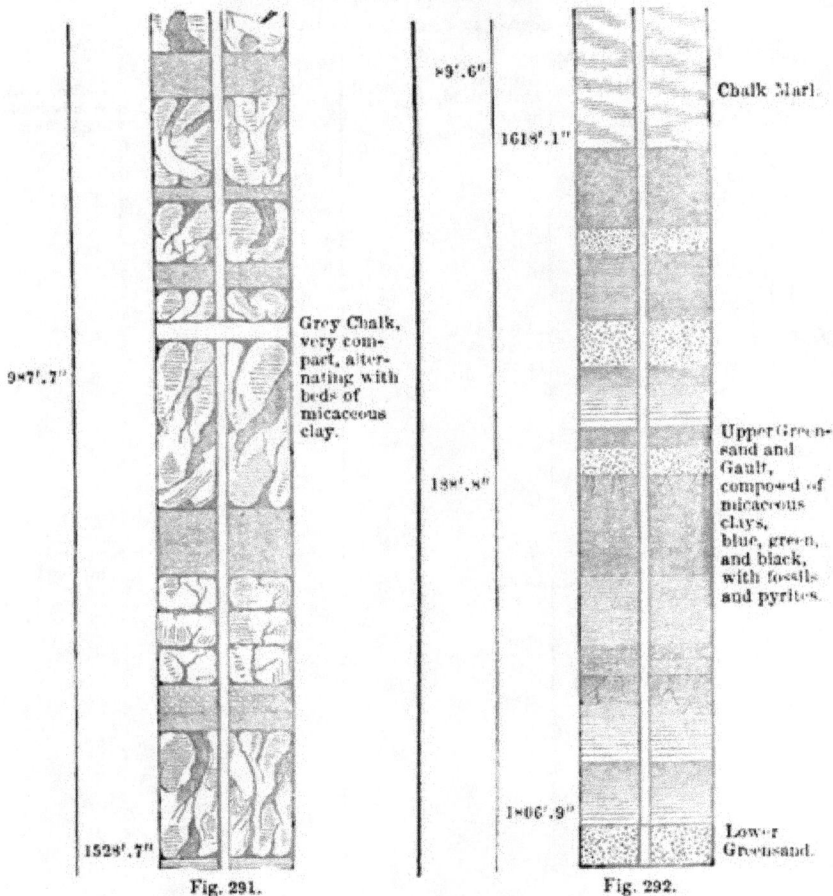

Fig. 291.

Fig. 292.

was decided to behave in a liberal manner, and to release him
from it, the town retaining his services for the completion of
the well, as also the right to use his patent machinery. The
difficulties encountered in carrying the excavation through the
clays of the upper strata were found to be so serious, that,

under the new arrangement, it required six years and nine
months of continuous efforts to reach the water-bearing stratum,
of which time the far larger portion was employed in travers-

BORING AT PASSY, PARIS.

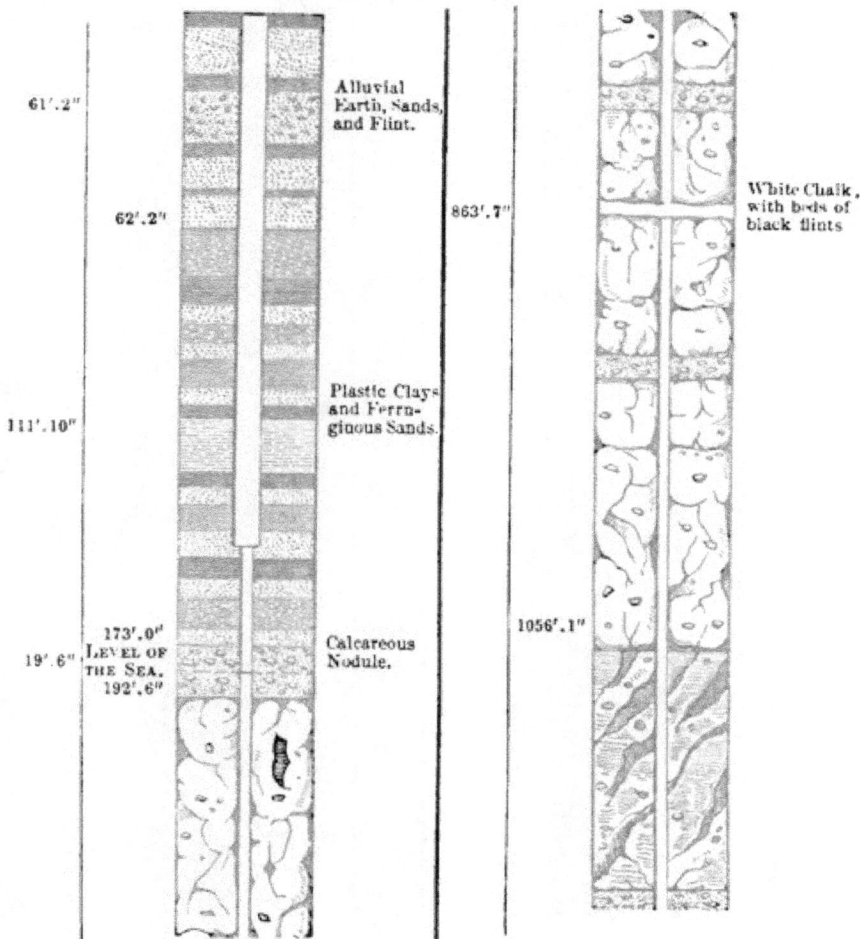

61′.2″

62′.2″

111′.10″

173′.0″
19′.6″ LEVEL OF
THE SEA.
192′.6″

Alluvial
Earth, Sands,
and Flint.

Plastic Clays
and Ferru-
ginous Sands.

Calcareous
Nodule.

863′.7″

1056′.1″

White Chalk,
with beds of
black flints

Fig. 293.　　　　　　　Fig. 294.

ing the clay beds. The upper part of this well was finally
lined with solid masonry, to the depth of 150 feet from the
surface; and beyond that depth tubing of wood and iron was

introduced. This tubing was continued to the depth of 1804 feet from the surface, and had at the bottom a length of copper pipe pierced with holes to allow the water to enter. At this

BORING AT PASSY, PARIS—*continued.*

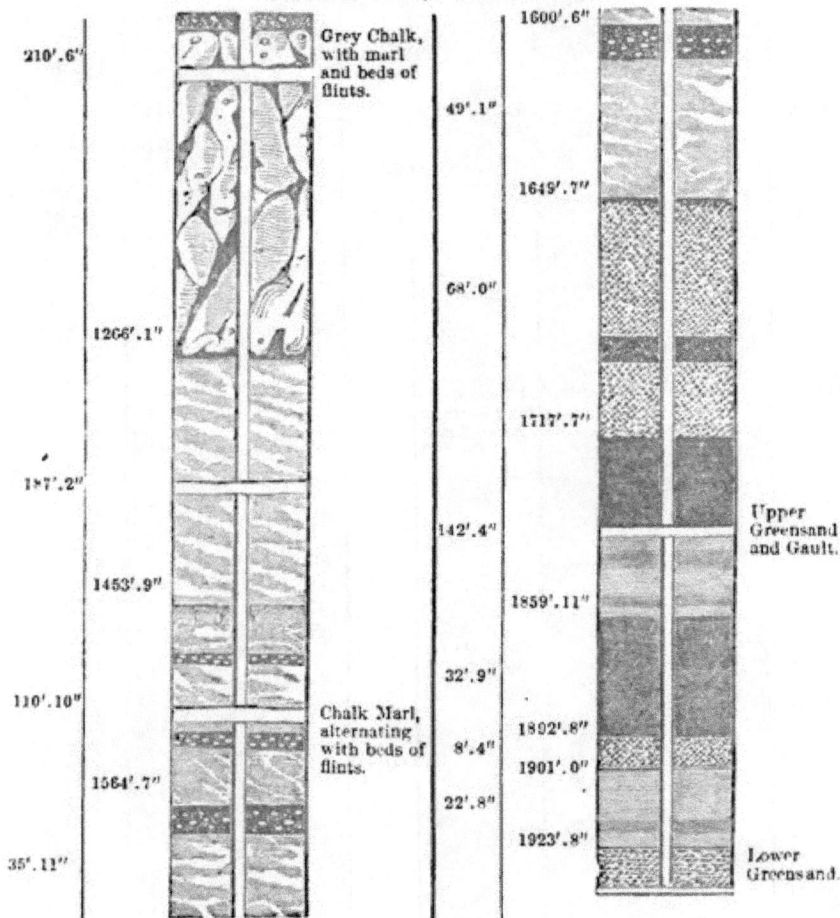

Grey Chalk, with marl and beds of flints.

Chalk Marl, alternating with beds of flints.

Upper Greensand and Gault.

Lower Greensand.

Fig. 295.　　　　Fig. 296.

depth the compound tubing could not be made to descend any lower; but the engineers employed by the city of Paris were convinced that they could obtain the water by means of a preliminary boring; and therefore they proceeded to sink in the interior of the above tube of 3·2809 feet diameter, an inner tube 2 feet 4 inches diameter, formed of wrought-iron plates 2 inches thick, so as to enable them to traverse the clays encountered at this zone. At last, the water-bearing strata were met with on the 24th of September, 1861, at the depth of 1913 feet 10 inches from the ground-line; the yield of the well being, at the first stroke of the tool that pierced the crust, 15,000 cubic mètres in 24 hours, or 3,349,200 gallons a day; it quickly rose to 25,000 cubic mètres, or 5,582,000 gallons a day; and as long as the column of water rose without any sensible diminution, it continued to deliver a uniform quantity of 17,000 mètres, or 3,795,000 gallons a day. The total cost of this well was more than 40,000l., instead of 12,000l., at which Kind had originally estimated it.

It may be questioned whether the engineers of the town were justified in passing the contract with Kind to finish the work within the time, and for the sum at which he undertook it; but they certainly treated him with kindness and consideration, in allowing him to conduct the work at the expense of the city of Paris for so long a period after the expiration of his contract. It seems, however, that the French well-borers could not at the time have attempted to continue the well upon any other system than that introduced by Kind; that is to say, upon the supposition that it should be completed of the dimensions originally undertaken. Experience has shown that both steining and tubing were badly executed at the well of Passy. The masonry lining was introduced after Kind's contract had expired, and when he had ceased to have the control of the works; the wrought-iron tubing at the lower part of the excavation being a subsequent idea. It has followed from this defective system of tubing—the wood necessarily yielding in the vertical joints—that the water in its upward passage escaped through the joints, and went to supply the basement beds of the Paris basin, which are as much resorted to as the London sand-beds for an Artesian supply; and, in fact, the level of the water has been raised in the neighbouring wells by the quantity let in from below, and the yield of the well itself has been proportionally diminished,

Fig. 297.

WELL AT PONDERS END.

PLAN

Fig. 298.

until it has fallen to 450,000 gallons a day. That the increased yield of the neighbouring wells is to be accounted for by the escape of the water from the Artesian boring is additionally proved by the temperature of the water in them; it is found to be nearly 82° Fahr., or nearly that observed in the water of Passy. This was an unfortunate complication of the bargain made between Kind and the Municipal Council, but it in no respect affects the choice of the boring machinery, which seems to have complied with all the conditions it was designed to meet. The descent of the tubes and their nature ought to have been the subject of special study by the engineers of the town, who should have known the nature of the strata to be traversed better than Kind could be supposed to do, and should have insisted upon the tubing being executed of cast or wrought iron, so as effectually to resist the passage of the water. At any rate, this precaution ought to have been taken in the portions of the well carried through the basement beds of the Paris basin, or through the lower members of the chalk and the upper green sand.

Ponders End.—Middlesex. At the works of the London Jute Company. It will be seen from the Figs. 297, 298, that this well is bored all but the top 4 feet, which is 5 feet across and steined with 9-inch work. The uppermost tube is 12 inches in diameter, decreased to 9 inches, and then to 8 inches, and ending with a 6-inch bore, unlined, in the chalk.

The strata passed were ;—

	Feet.	In.
ALLUVIUM, 6 feet ;—		
Clay and Mud	3	6
Peat	2	6
SAND AND SHINGLE GRAVEL	7	0
LONDON CLAY, 15 feet ;—		
Blue Clay	8	0
Sandy Clay (basement bed ?)	7	0
READING BEDS, 49½ feet ;—		
Dead Sand	10	0
Mottled Clays	22	0
Sand and Metal (pyrites ?)	1	0
Sandy Clay	3	0
Sand and Pebbles	4	0
Dead Sand	1	6
Dead Sand and Pebbles	1	0
Sand and Pebbles	7	0
THANET SAND (?), 35 feet ;—		
Green Sand	27	0
Dead Sand	8	0
To Chalk	112	6
IN CHALK	290	6
Total	403	0

The water at this well overflows.

Hampstead.—Middlesex. Well at the Brewery. Consists of a shaft 340 feet deep continued by a bore-hole 5 inches in diameter into the chalk. Water level about 320 feet from surface.

	Feet.	In.
Section ;—		
Made earth	6	0
LONDON CLAY ; —		
Clay, with Shells	134	0
Hard Clay and Nodules of Spar	2	0
Clay	54	0
Clay, with shells and pyrites	10	0
Blue Clay	202	0
WOOLWICH BEDS ; —		
Clay and Pebbles	5	0
Clay and Sand	18	0
Grey Sand	12	0
Flints	2	0
Chalk	155	0
Total	600	0

WELL AT MEUX' BREWERY.

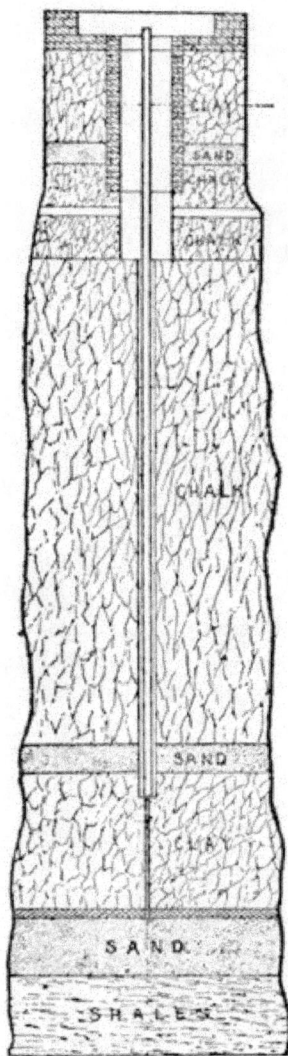

VERTICAL SCALE.

Feet 0   50   100   150   200

Fig. 299.

Stowmarket. — Suffolk. Well at Hewitt's Mill.   Yield abundant.

STRATA ;—	Feet.
Loam, Sand and Gravel	20
Sharp Sandstone	80
Chalk and Flints	200
Total	300

Tottenham Court Road.—Middlesex. Well at Meux' Brewery, Fig. 299, yield at 1022 feet from surface, or 21 feet in the lower greensand, 1500 gallons an hour.

STRATA :—	Feet.	In.
London Clay and Tertiaries ..	156	0
Chalk, with flints	347	0
Chalk, without flints	305	0
Upper Greensand	28	0
Gault	160	0
Coprolites	0	6
Limestone	4	6
Lower Greensand	66	0
Mottled Red and Green Argilla- ceous and Micaceous Shales, palæozoic	77	0
Total	1144	0

Bognor.—Isle of Wight. Well at Waterworks has a shaft lined with 9-inch brickwork for 80 feet, continued by a bore-hole to a total depth of 330 feet. The yield at 80 feet from surface is 150,000 gallons a day.

STRATA ;—	Feet.
Brick earth, running sand and clay	58
Sand	22
Red and Blue Clay	34
Chalk	216
Total	330

Freshwater.—Isle of Wight. Well, Figs. 300, 301, sunk at Golden Hill for H.M. Government. The diameter of the shaft

WELL AT FRESHWATER, ISLE OF WIGHT.

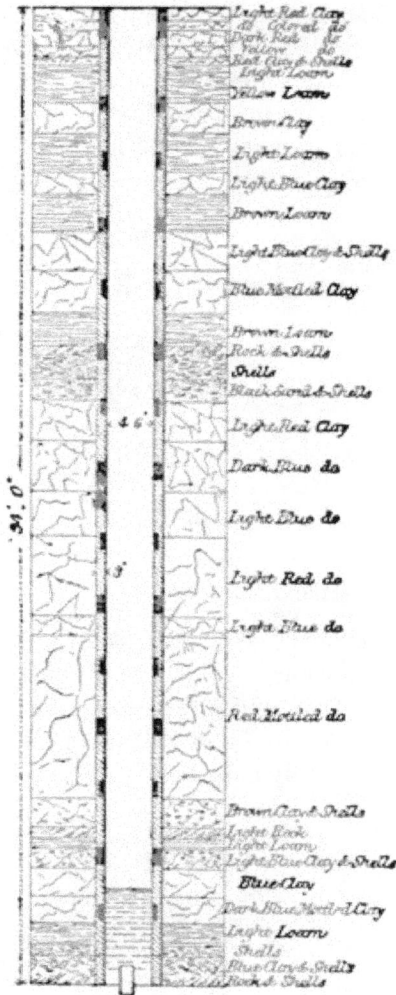

Fig. 300.

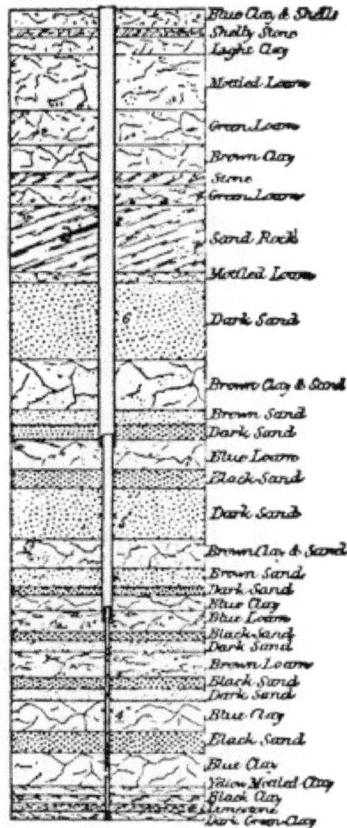

Fig. 301.

is 4 feet 6 inches, brickwork 9 inches thick, there are 3 feet in cement at the top of the well, and 3 feet 9 inches at the bottom. There are four courses in cement every 5 feet, internal work four courses in cement every 10 feet. The bore-

Fig. 302.

WELL AT WINCHFIELD,
HANTS.

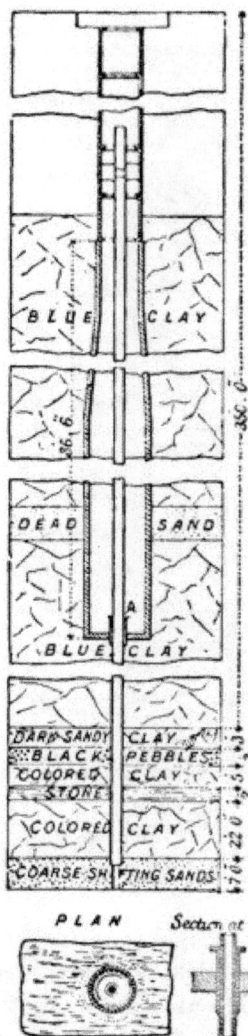

PLAN　　　Section at A

Fig. 303.　　Fig. 304.

hole is lined throughout with pipes of 6 inches, 5 inches, and 4 inches diameter respectively.

Winchfield, Hants.—Well, Figs. 302 to 304, at the brewery of Messrs. W. Cave and Son. The shaft above the steining is lined with iron cylinders into which the bore-pipe is carried up.

The strata passed were ;—

	Feet.
Made Earth, Soil, Gravel, Blue Clay and Dead Sand　..　..	350
Dark Sandy Clay　..　..　..	3
Black Pebbles　..　..　..　..	2
Coloured Clay　..　..　..　..	5
Stone (septaria ?)　..　..　..	2
Coloured Clay　..　..　..　..	22
Coarse Shifting Sands　..　..	7
Total　..　..　..　..	391

MISCELLANEOUS.

Ulster.—Ireland. Well at Ross & Co.'s Mineral Water Factory, has a shaft lined with iron cylinders 70 feet deep, continued to a total depth of 226 feet by a bore-hole. Supply abundant.

STRATA :—

	Feet.
Made Earth　..　..　..　..	5
Silt Blue Clay, with shells　..	21
Gravel with prehistoric remains	7
Stiff Red Clay ..　..　..　..	37
Gravel　..　..　..　..　..	8
Red Sandstone ..　..　..　..	146
Fine Gravel　..　..　..　..	2
Total　..　..　..	226

Bourne, Lincolnshire.—The boring, 4 inches in diameter, passed through oolitic strata to a depth of 92 feet. Below the alluvial gravel and alluvion a hard shelly limestone, 32 feet in thickness, was encountered. The bore-hole here was made slightly conical, to admit of the taper end of a cast-iron pipe being inserted and driven tightly, to exclude any surface water, and to prevent water from the bore escaping into the gravel, and thus losing its full power to rise above the surface. The boring was then continued through various beds until it reached a stratum, 6 feet thick, of compact hard rock; in passing through which, at 92 feet below the surface, the tool fell suddenly about 2 feet, evidently into a chasm or hollow, striking upon the hard surface of the underlying rock. The water immediately rushed up with great force, and drove the men from their work; and it was not without difficulty that the joints for attaching a curved pipe and sluice valve at the surface could be accomplished. The water rose to 39 feet 9 inches above the ground; the yield at the surface level is at the rate of 567,000 gallons a day.

Particulars of Wells in the London Basin.

The following Table, compiled from the Government Memoirs and other reliable sources, furnishes in a condensed form the most important particulars relating to wells, and trial bore holes, comprised within the geographical area known as the London Basin.

The first column gives the name of the place where the well is situated, the second column that of the county, and the third column the precise locality; the sixth column, in cases where the well passes *through* the tertiaries, is the depth to the chalk. The following abbreviations have been employed: B. for Bedfordshire; Berks, Berkshire; Bucks, Buckinghamshire; E., Essex; H., Hampshire; Herts, Hertfordshire; K., Kent; M., Middlesex; S., Surrey.

O.D. stands for, above Ordnance Datum; T., above Trinity high-water mark.

Place			feet.	feet.	feet.	feet.	feet.	Remarks
Abridge	E.		100	190	290	—	30	London Clay, 280 feet.
Acton	M.	Brewery	—	—	284	119	12	
Ditto	"	Mr. Engleheart's	—	—	315	135	40	
Ditto, East	"	Mr. Wood's	—	—	267	68	—	
Albany Street	"	Mr. Davis's	—	—	182	125	—	100 feet O.D.
Aldershot Place	H.	London	—	—	194	69½	—	260 feet O.D.
Ditto	"		—	—	148	—	—	245 feet O.D.
Amwell End	Herts.	New River Company	72	317¾	36	383¾	—	Yield about 2,500,000 gallons a day.
Amwell Hill	"		90	70	2	158	—	
Amwell Marsh	"		99	283	4	378	—	
Arlesey	B.	Asylum	100	365	7	120	—	283 feet O.D.; water rises into shaft; yield 2640 gallons an hour.
Ash	S.	S.W. Railway Station	137	600	370	230	88	290 feet O.D.
Bank of England	M.	London	123	197½	231½	100	—	About 27 feet T.; yield, 35 gallons a minute.
Bagshot	S.	Orphan Asylum	140	523	646	—	—	Last 192 feet London Clay.
Balham Hill	"	Near Clapham Common	122	347	317	—	—	Last 40 feet Thanet sands.
Barking	E.	Byfron's	—	270	140	230	30	Bottom in hard pebbles.
Barnet, East	Herts.	Lion's Down	—	—	162	280	130	Shaft half stained, half iron cylinders.
Ditto, New	S.	Near Railway Station	137	302	159	—	130	
Battersea	S.	Jones's Works	249	—	249	—	—	Yield said to equal 15,000 gallons a day.
Ditto	"	Beaufoy's Works	210	—	210	—	—	

Bethnal Green	M.	Workhouse	—	—	190	15	118	
Bearwood	Berks.	Mr. Walter's	—	—	350	57	80	
Beaumont Green	Herts.	Near Cheshunt	183½	—	126½	118	—	
Belleisle	M.	Pashes and Co.'s	—	—	185	92	—	
Berkeley Square	S.	London	160	156	221	—	—	9 feet O.D.; yield plentiful.
Bermondsey	"	Crimscott Street	—	232	120	110½	16	Yield 30 gallons a minute.
Ditto	"	Donkin's Works	—	20	91½	—	8	
Berry Green	Herts.	Hadham	40	110	60	8	60	
Bexley	K.	Brickfield	65	110	129½	45½	140	Supply 10,000 gallons a minute.
Bishop Stortford	Herts.	Waterworks	160	125	116½	183¼	78	Good supply.
Ditto	"	Hockerill	85	—	90	120	—	
Ditto	"	New Road	77	—	56	21	—	
Blackfriars	M.	Apothecaries' Hall	—	—	218	76	—	
Blackheath	K.	Near Enfield Terrace	—	250	109	30	—	
Bognor	H.	Waterworks	80	—	114	216	80	Yield 150,000 gallons a day.
Boston Heath	K.	Near Woolwich	—	118	130	70	—	
Bow	M.	Starch Works	176	213½	174	150	—	
Boxley Wood	K.	Near Maidstone	386½		3	600	—	382 feet T.; last 78½ feet in chalk, marl, and gault.
Braintree	E.	Near Pod's Brook	55	190	228	17	216	Yield 11,500 gallons an hour.
Brentford	M.	Brewery	30	338	315	53	5	
Broad Mead	"	New River Company	25	533	16	512	—	
Bromley	K.	Gas Works	50	120	150	20	—	Supply abundant.
Ditto	"	Widmore Kiln	52	98	140	10	61	
Ditto	"	Ditto	55	85	120	10	—	
Ditto	"	Tylney Road	77	85	137	25	—	
Ditto	"	Waterworks	—	—	70	180	—	Yield 500 to 600 gallons a minute.
Broxbourne	Herts.	—	81	—	81	6	—	
Bushey	"	Near Watford	142	24	145	21	—	Water overflowed.

PARTICULARS OF WELLS—*continued.*

Name of Place.	County.	Locality.	Depth of Shaft.	Depth of Bore.	in Tertiary Strata.	in Chalk.	to Water from Surface.	Remarks.
			feet.	feet.	feet.	feet.	feet.	
Camberwell	S.	The Grove	180	—	208	300½	90	100 feet O.D.
Camden Station	M.	L. and N. W. Railway	—	220	234	166	150	Good supply.
Camden Town	M.	Pickford's	—	—	215	82	120	
Ditto	M.	Whitaker's Brewery	235	75	210	90	190	
Canterbury	K.	Orphan Asylum	—	—	145	—	120	709 feet T.; through chalk, and 39 feet into upper greensand.
Caterham	S.	Waterworks	—	—	89	349	—	
Chelmsford	E.	Moulsham	200	368	366	202	76	Water overflowed at first.
Cheshunt	Herts.	New River Company	14½	27	107½	63½	120	Yield, 702,000 gallons a day.
Ditto	"	Theobald's Park	71	131½	121½	81	65	
Chiswell Street	M.	Whitbread's Brewery	183	150	183	150	132	
Chiswick	"	Griffin Brewery	204	200	297	107	—	Yield, 14 gallons a minute.
Ditto	"	Lamb Brewery	203	194	293	104		
Ditto	"	Ditto	8	339	237	50		
Clewer Green	Berks	Capt. Winterbottom's	42	294	270	63		Water found at 203 feet down.
Ditto	"	Wycombe Cottage	20	216	169	97		
Colnbrook	M.	Paper Mills	—	—	207	175	—	
Colney Hatch	"	Asylum	137	193	189	141	120	70 feet O.D.
Covent Garden	"	Market	140	218	260	98	110	157 feet T.
Cricklewood	"	Near Hampstead	225	85	221	19		
Crosness	E.	Pumping Station	—	—	179	51	230	Yield, 52,000 gallons in 24 hours.

								Remarks
Croydon	S.	Well for Local Board	77	—	11	62	—	Yield, 1,500,000 gallons a day.
Ditto	„	New Well	—	—	15	137	11¼	
Ditto	„	Sewage Works	—	—	17	—	—	
Dartford Creek	K.	Paper Mills	31	49	33	50	—	Supply good.
Ditto	„	Ditto	10	240½	30	220½	2	
Denham	Bucks.	Tile House	110	83	67	128	85	
Deptford	K.	Waterworks	27	—	14	13	—	
Dulwich	S.	Champion Hill	—	—	210	298	—	20 feet O.D.
East Ham Level	E.	Beckton Gas Works	25	175	117	83	2	
Edgware	M.	Mr. Day's	—	—	290	45	40	
Edgware Road	„	The Hyde	—	—	101	57	70	
Edlesborough	Bucks.	Well, near Mill	—	301	—	—	—	6-inch bore; through 50 feet of chalk marl to lower greensand.
Eltham	K.	Dr. King's	46	—	46	10	17	
Ditto	„	The Moat	110	123	100	41½	25	
Ditto	„	Mr. Tuck's	44	107	122½	3	170	
Ditto	„	Well Hall	—	—	104	94	4	
Ditto Park	„		—	—	122	132	260	Slow spring.
Enfield Lock	E.	Small Arms Factory	45	235¾	155¼	4	243	
Epping	„	Waterworks	275	129	400	20	—	
Erith	K.	Mineral Oil Company	166	—	146	96	—	
Farnham	S.	Near Hale Farm	176	—	80	—	—	
Finchley	M.	Nether Street	247	340	284	303	—	
Fleet Street	„	London, Shoe Lane	100	225	100	225	—	
Freshwater	H.	Golden Hill	94	86	180	—	—	
Fulmer	Bucks.	J. Kay's	85	—	47¾	37¾	—	Through gravel and Reading beds.
Golden Lane	M.	Baths and Washhouses	158	—	151½	6¾	—	65 feet O.D.
Gravesend	K.	Church Street	10	234	120	121	8	Supply good and abundant.
Greenwich	„	Brewery	22	158	80	100	11	

B

PARTICULARS OF WELLS—*continued.*

Name of Place.	County.	Locality.	Depth of Shaft.	Depth of Bore.	in Tertiary Strata.	in Chalk.	to Water from Surface.	Remarks.
			feet.	feet.	feet.	feet.	feet.	
Greenwich	K.	East Street	189	—	159	30	19	7 feet T.; supply 120 gallons a minute.
Ditto	"	Hospital Brewery	155	150	121½	180½		
Hackney Road	M.	Wiltshire Brewery	96	315½	152½	259	80	
Haggerstone	"	Imperial Gas Works	118¾	302	164½	256		
Hainault Forest	E.		165	—	110	55		
Halstead	"	The White Hart	—	—	170	30		Yield, 16 gallons a minute.
Hammersmith	M.	Average of four wells	—	—	245	68	—	Now not used.
Hampstead	"	Lower Heath	320	130	378	72	320	
Ditto	"	Brewery	34	566	415	155	147	77 feet T.
Hampstead Road	"	Eagle Brewery	138	94	146	86	106	Water to surface.
Ditto	"	Reservoir	244		152	92		
Hanwell	"	Asylun	230	90	290	30	125	226 feet O.D.
Harrow	"	Waterworks	192¾	219	158½	254	196	176 feet O.D.
Haverstock Hill	"	Orphan School	250	160	312	78	27	
Hayes	"	Dawley Court	19	300	231	88	76	
Hendon	"	Mr. Booth's	—	—	244	132	6	Yield, plentiful.
Hertford	Herts.	Corporation Water-works	—	81	12	69		
Highbury	M.	Brewery	101	210	180	134	95	Yield, 1000 gallons an hour.
Ditto	"	New Park	136	113	199	50	2	
Hoddesdon	Herts.		52	234	21½	261½		
Holloway	M.	New River Company	149	260	240	100		

Ditto	,,	City Prison	—	—	217	102		
Ditto	,,	Hanley Road	—	—	67	13		
Ditto	,,	Red Cap Lane	—	—	210	90		
Ditto	,,	Islington Workhouse	231	314	300	218		
Hornsey	,,	Near Church	—	—	202	48		
Ditto	,,	The Priory	—	—	225	—	70	
Horselydown ..	S.	Anchor Brewery ..	100	162	158	104	50	
Hoxton	M.	— —	152	10	151	11		
Hyde Park Corner	,,	St. George's Hospital	200	137¼	319¼	18	100	50 feet O.D.; yield, 3300 gallons an hour.
Ickenham	,,	Public Well	61	80	64	80		
Isle of Dogs	,,	Oil Mills	27	337	121½	239½	10	
Isle of Grain ..	K.	Fort	180	140	320	—	20	21 feet O.D.
Isleworth	M.	Sion House	—	—	420	115	—	Water overflowed at the rate of 5 gallons a minute.
Ditto	,,	Mr. Wilmot's	—	327	327	—	—	Water rose above surface.
Islington Green ..	,,	Webb's Mineral Water Works	—	320	176	144	200	
Kensington	,,	Brewery	—	—	197	—	—	16 feet T.
Ditto	,,	Britannia Brewery ..	100	170	270	—	88	
Ditto	,,	Horticultural Society	200	201	317	84	100	60 feet O.D.
Ditto	,,	Workhouse	—	—	270	100		
Ditto Gardens	,,	Serpentine	263	58	263¼	57¾	105	60 feet O.D.; yield, 250 gallons a minute.
Kentish Town ..	,,	Waterworks	539	763	321½	644¾	—	Through London clay, 236 feet; London tertiaries, 88½ feet; chalk, 644¾ feet; upper greensand, 13¾ feet; gault, 130½ feet; and into lower greensand (?), 188½ feet.

PARTICULARS OF WELLS—continued.

Name of Place.	County.	Locality.	Depth				to Water from Surface.	Remarks.
			of Shaft.	of Bore.	in Tertiary Strata.	in Chalk.		
			feet.	feet.	feet.	feet.	feet.	
Kilburn	M.	Brewery	250	30	235	45	150	25 feet O.D.; yield, about 44,000 gallons a day.
Kingsbury ..	S.	Brent Reservoir ..	101	139	132	108	—	
Kingston-on-Thames	S.	Brook Street ..	90	360	371	99	50	
Knightsbridge ..	M.	—	240	—	240	—	—	
Lambeth ..	S.	Beaufoy's Vinegar Works	100	275	201	174		Yield 32 gallons a minute.
Ditto	„	South Lambeth Road	25	166	187	4		
Ditto	„	Bethlehem Hospital ..	30	161	191	20	15	
Ditto	„	Lion Brewery, Belvedere Road	—	—	245	173	40	
Ditto	„	Duke Street, Clowes & Sons'	26	184	210	—		
Lea Bridge	M.	Waterworks	118	—	100	18		
Leicester Square ..	„	Alhambra	150	195	244	101		
Limehouse	„	Johnson's, Commercial Road	90	110	190	10		
Ditto	„	Brewery, Fore Street	—	—	130½	—		
Ditto	„	Brewery, Narrow St.	143	153	141	155		
Liquorpond Street	„	Reid's Brewery	222½	40	136	126½	121	70 feet O.D.; yield, 277,200 gallons in 24 hours.
Long Acre	„	Combe & Co.'s Brewery	263	228	223	268	—	70 feet O.D.; yield, 90 gallons a minute.

								Remarks
Loughton	E.	Railway Station	—	535	324	211	90	No water from chalk.
Ditto	,,	—	—	1092	213	648	—	Said to end in lower green-sand.
Lower Morden	S.	On the Green	20	365	310	45	—	Water to surface.
Luton	B.	Waterworks	50	272	—	322	—	
Maldon	E.	Waterworks	234	—	234	—	—	Entirely through London clay.
Margate	K.	Cobb's Brewery	31	213	—	374	156	
Marylebone Road	M.	London; a Brewery	186	101	232	53	—	
Mile End	,,	Mann's Brewery	195	—	185	605	—	Said to end in greensand; yield good.
Ditto	,,	Charrington's Brewery	204	—	202	2	103	33½ feet T.; yield, 60,000 to 70,000 gallons a day.
Mile End Road	,,	City of London Union	—	—	175	10	70	Level of T.
Millbank	,,	Distillery	115	190	205	100	—	5½ feet T.
Ditto	,,	Westminster Brewery	—	—	225	70	—	
Mitcham	S.	Nightingale's Factory	—	211	189	22	—	
Monkham Park	E.	Near Waltham Abbey	225	125	304	76	50	
Mortlake	S.	Mortlake Brewery	30	288	287	31	50	Yield, 14,000 gallons a day.
Ditto	,,	Mr. Randell's	—	365	315	50	60	
New Cross	K.	Naval School	50	130	125	55	4	
Northolt	M.	Near Harrow	12	228	180	60	—	
Notting Dale	,,	Near Notting Hill	—	—	214	12	—	
Notting Hill	,,	Mr. Knight's	—	—	230	200	—	
Old Kent Road	S.	Welsh Ale Brewery	—	—	30	170	—	
Old Windsor	Berks.	Pelham Place	—	180	222	9	—	
Ditto	M.	The Union	60	—	240	47	—	10 feet O.D.
Orange Street		Back of National Gallery	174	126	250	50	115	
Oxford Street	,,	Star Brewery	166	170	158	178	—	42 feet T.
Peckham	S.	Marlborough House	—	—	100	123		

PARTICULARS OF WELLS—*continued.*

Name of Place.	County.	Locality.	Depth				to Water from Surface.	Remarks.
			of Shaft.	of Bore.	in Tertiary Strata.	in Chalk.		
			feet.	feet.	feet.	feet.	feet.	
Penge	S.	Palace Grounds	250	310	358	202	90	
Pentonville	M.	Brewery, Caledonian Road	219½	—	219½	45	180	To chalk.
Ditto	"	Prison	170	200½	219½	151	—	
Ditto	"	Cubitt's Works	188	368	188	127	—	2 feet T.
Ditto	"	Brewer Street	30	—	271	100	36	1 foot T.
Ditto	"	Simpson's Factory	—	—	231	80	—	
Pinner	"	Hatch End	140	—	60	128	—	
Plaistow	E.	Odam's Manure Works	4	399	170½	200½	—	Water overflows.
Ponders End	M.	London Jute Company	20	42	112½	—	—	
Ditto	"	Crape Works	—	—	62	96½	—	
Ditto	"	Local Board (Speller)	23	181	106	107	—	43 feet T.
Ditto	"	Waterworks	297	—	97	—	—	Water abundant and good.
Purley Hall	E.	Near Catewdon	16	—	257	200	—	
Ratcliffe	M.	Queen's Head Brewery	—	236	160	102	—	
Ditto	"	Marine Brewery	—	137	150	—	—	
Ditto	"	Ravenhill's	150	100	171	79	80	
Regent's Park	"	Colosseum	183	—	184	215	120	
Ditto	"	Mr. Day's	—	91	224	50	—	Yield, 90,000 gallons a day.
Ditto	"	Zoological Gardens	—	—	276	163	—	
Richmond	S.	Old Waterworks	155	—	416	76	—	
Ditto	"	Star and Garter	—	—	145	10	—	
Romford	E.	Ind, Coope, & Co.	—	—	—	—	—	

Locality		Name	Dist.						Remarks
Rotherhithe .. :		Brandram's Works ..	S.	30	222	107	145	27	Yield, 100,000 gallons in 12 hours.
Ditto .. :		Tunnel Flour Mills ..	"	—	—	125	135	—	15 feet O.D.; yield, 80 gallons a minute.
Ruislip .. :		Near "The George" ..	M.	15	90¾	73	30	—	Water to surface.
Saffron Walden :			E.	—	—	—	1000	—	
Sandhurst .. :		Well at College ..	Berks.	70	603	62	8	20	Trial boring; chalk reached.
Sandwich .. :		The Bank ..	K.	300	81	384	—	—	
Sheerness .. :		Waterworks ..	"	230	125	455	331	53	5½ feet O.D.; yield, 10,000 gallons an hour.
Ditto .. :		Dockyard ..	"	300	230	199	314½	120	Yield, 675 gallons an hour.
Shoreditch .. :		Truman's Brewery ..	M.	112	150	77½	100	61	Yield, 7¼ gallons a minute.
Shorne Meade Fort :		Near Gravesend ..	K.	59	103	189	21	7	Yield, 1000 gallons an hour.
Shortlands .. :		Near Bromley ..	"	28	—	107	17½	70	
Slough .. :		Eton Union ..	Bucks.	—	—	94	170½	100	
Ditto .. :		Royal Nursery ..	"	—	—	102½	260	—	
Ditto .. :		Upton Park ..	"	117	233	90	70	—	Heading into chalk.
Ditto .. :		Waterworks ..	"	417	—	230	92	—	Old well.
Smithfield .. :		Booth's Distillery ..	M.	142	158	417	211	81	Level of T.; yield, 300 gallons a minute.
Southend .. :		Waterworks ..	E.	115	288	208	109	—	2 feet T.; yield, 33 gallons a minute.
Southgate .. :		Betstile (N. R. Co.) ..	M.	132	173	212	151	46	Water to surface.
Southwark .. :		Barclay's Brewery ..	S.	63	—	196	30	—	
Ditto .. :		Guy's Hospital ..	"	100	210	369	100	—	
Staines .. :		Ashby's Brewery ..	M.	25	186	33	154	—	Yield, 33 gallons a minute.
Stifford .. :		S.E. of Church ..	E.	56	314	210	100	—	
Stockwell Green :		Walthum's Brewery ..	S.	—	—	211	294	—	Yield, 46 gallons a minute.
Ditto .. :		Hammerton's Brewery ..	"	—	—	106	—	—	
Stratford .. :		Great Eastern Works	E.	—	—	—	—	—	

PARTICULARS OF WELLS—*continued.*

Name of Place.	County.	Locality.	Depth					Remarks.
			of Shaft.	of Bore.	In Tertiary Strata.	In Chalk.	to Water from Surface.	
			feet.	feet.	feet.	feet.	feet.	
Stratford	E.	Savill Bros.' Brewery	112½	—	109½	3	—	Supply abundant.
Ditto	"	Langthorn Chemical Works	60	395	132	323	—	
Streatham	S.	The Common	100	185	285		—	
Sudbury	M.	London and North-Western Rail. Station	200	—	120	80	—	
Tottenham	"	Warne's Works	—	—	147	101	—	
Ditto	"	Long Water	—	—	149½	101½	—	
Ditto	"	Tottenham Hall	—	253	153	100	—	
Tottenham Court Road	"	Meux's Brewery	230	894	156	652	—	85 feet O.D.; is through chalk, greensand and gault into palæozoic strata; yield 25 gallons a minute.
Tower Hill	"	Royal Mint	195½	202	195½	202	80	Yield, 450 gallons a minute.
Trafalgar Square	"	London	168	228	248	148	—	
Turnford	K.	New River Company	176	604	99	681	—	
Upchurch	"	Burntwick Island	—	256	233	91	—	Good supply at bottom.
Ditto	M.	Milford Hope Marshes	90	204	210	245	10	
UpperThamesStreet	"	City of London Brewery	121	415	210	38½	3	
Uxbridge	"	The Dolphin	—	—	81½		15½	
Ditto	"	Near Market Place	—	—	104	28		

Parish	Co.	Well						Remarks
Ditto	"	Town Well	—	—	109	30	19	
Ditto	"	Page's Lane	98	—	98	—	19	To chalk.
Ditto	"	Near "King's Arms"	21	84	108	—	51	
Ditto	"	New Year's Green Farm	63	38½	63	39½	39	
Ditto	"	Hurdle Yard	78	104½	78	36	29	
Ditto	"	Near Meeting House	41½	152	115	38	—	
Ditto	"	The Union	51	600	175	66	—	
Vange	E.	Near Pitsea	140	186	521	102	39	London Clay, 395 feet
Vauxhall	S.	Burnett's Distillery	11	186	224	192	29	Yield, 80 gallons a minute.
Ditto	"	Bond Street	161	270	78	4	55	
Waltham Abbey	E.	Brewery	—	—	160	—	52	Water supply from bed of sand.
Walthamstow Marsh	"	East London Waterworks	170	164	152	140	—	15 feet T.
Wandsworth	S.	Young & Bainbridge's	—	—	274	60	45	Yield, 10 gallons a minute.
Ditto	"	Prison	—	—	357	126½	80	Yield, 27 gallons a minute.
Ditto	"	County Asylum	210	150	331	6	30	
Watford	Herts	Waterworks	12	67	17	133	—	Yield 200 gallons a minute.
Westbourne Grove	M.	Hippodrome	—	274	309	7	—	
West Drayton	"	Victoria Oil Mills	3	—	186	100	—	Water overflowed.
Ditto	"	Vitriol Works	—	146	133½	45½	—	
Ditto	"	Drayton Mills	—	—	149	—	—	
West Ham	E.	Mr. Tucker's	—	—	132	306	—	To chalk.
Ditto	"	Union	—	—	110	55	—	
West India Dock	M.	South of Export Dock	—	—	120	210	70	
Westminster	"	Artillery Brewery	—	—	230	—	60	
Ditto	"	Chartered Gas Works	—	184	225	51	—	
Ditto	"	Vickers' Distillery	116	—	249	—	—	Yield, 94 gallons a minute.
Ditto	"	Swallow Street	—	—	210	—	—	

PARTICULARS OF WELLS—*continued.*

Name of Place.	County.	Locality.	Depth of Shaft.	of Bore.	In Tertiary Strata.	In Chalk.	to Water from Surface.	Remarks.
			feet.	feet.	feet.	feet.	feet.	
Whitechapel	M.	Furze's Brewery	130	218	218	100	85	
Ditto	"	Smith, Druce, & Co.'s	111½	—	141½	—	35	39 feet T.
Ditto	"	Smith's Distillery	166	264	210	160	36	36 feet T.
Willesden	"	Mr. Kilsby's	—	—	273	57	39	
Wimbledon	S.	Convalescent Hospital	200	367	557	30	50	
Ditto, New	"	Opposite "White Hart"	—	—	193	75		
Winchfield	H.	Cave's Brewery	—	—	331	—		
Windsor	Berks.	Ower Lodge	40	175	173	40	—	
Ditto	"	Royal Brewery	72	—	72	—	—	Through clay and running sand to chalk.
Ditto	"	Jennings' Brewery	—	—	30	560	12	
Winkfield Plain	"	Captain Forbes'	—	—	304	125	70	
Witham	E.	—	—	—	306	—	5	
Woodley Lodge	Berks.	3 miles east of Reading	95	33	130	—		
Woolwich	K.	Well of Arsenal	—	550	54?	311½	37	Yield 650 gallons a minute.
Ditto	"	Paper Factory	—	—	52	311½		Yield good.
Ditto	"	Dockyard	—	608	20	388	70	Water overflows.
Wormley	Herts.	Nunsbury	26	76½	80½	22	62½	
Ditto	"	West End	85	150½	72	62½	5	
Wormwood Scrubbs	M.	—	—	—	250	116		

CHAPTER IX.

TABLES AND MISCELLANEOUS INFORMATION.

The following tabulated form shows the order of succession of the various stratified rocks with their usual thicknesses.

Groups.				Strata.	Thickness in Feet.
CAINOZOIC, OR TERTIARY	RECENT			1 Modern Deposits.	
	PLEISTOCENE			2 Drift and Gravel Beds	20 to 100
	PLIOCENE			3 Mammaliferous Crag	10 to 40
				4 Red Crag	30
				5 Suffolk (Coralline) Crag	30
	MIOCENE			6 Faluns (Touraine) Molasse Sandstones	6000
	EOCENE	UPPER		7 Hempstead Series	170
				8 Bembridge Series	110
				9 Headon Series	200
		MIDDLE		10 Barton Beds	300
				11 Bagshot and Bracklesham Series	1200
		LOWER		12 London Clay and Bognor Beds	200 to 520
				13 Woolwich Beds & Thanet Sands	100
MESOZOIC, OR SECONDARY	CRETACEOUS			14 Maestricht Beds	110
				15 Upper Chalk	300
				16 Lower Chalk and Chalk Marl	400
				17 Upper Greensand	130
				18 Gault	100
				19 Specton Clay	130
				20 Lower Greensand	250
	WEALDEN			21 Weald Clay	150
				22 Hastings Sands	600
	JURASSIC	PURBECK		23 Purbeck Beds	150
		UPPER OOLITE		24 Portland Rock and Sand	150
				25 Kimmeridge Clay	400
		MIDDLE OOLITE		26 Upper Calcareous Grit	40
				27 Coralline Oolite	30
				28 Lower Calcareous Grit	40
				29 Oxford Clay	400
				30 Kellaways Rock	30

Groups.	Strata.	Thickness in Feet.
MESOZOIC—continued. JURASSIC — LOWER OOLITE	31 Cornbrash	10
	32 Forest Marble and Bradford Clay	50
	33 Great Oolite	120
	34 Stonefield Slate	9
	35 Fuller's Earth	50 to 150
	36 Inferior Oolite	80 to 250
LIAS	37 Upper Lias Shale	50 to 300
	38 Marlstone and Shale	30 to 200
	39 Lower Lias and Bone Beds	100 to 300
TRIASSIC or NEW RED SANDSTONE	40 Variegated Marls or Keuper	800
	41 Muschelkalk.	
	42 Red Sandstone or Bunter	600
PALÆOZOIC, or PRIMARY. PERMIAN or MAGNESIAN LIMESTONE	43 Red Sand and Marl	50
	44 Magnesian Limestone	300
	45 Marl Slate	60
	46 Lower Red Sandstone	200
CARBONIFEROUS	47 COAL MEASURES	3000 to 12,000
	48 Millstone Grit	600
	49 Mountain Limestone	500 to 1400
	50 Limestone Shales	1000
DEVONIAN or OLD RED SANDSTONE	51 Upper Devonian	
	52 Middle Devonian	3000 to 8000
	53 Lower Devonian and Tilestone	
SILURIAN — UPPER	54 Ludlow Rocks	2000
	55 Wenlock Beds	1800
	56 Woolhope Series	3050
MIDDLE	57 Llandovery Rocks	2000
	58 Caradoc and Bala Rocks	5000
LOWER	59 Llandeilo Rocks	4000
	60 Lingula Flags	8000
CAMBRIAN	61 Longmynd and Cambrian Rocks	20,000
AZOIC. METAMORPHIC	Clay Slate, Mica-Schist. Gneiss, Quartz Rocks.	
IGNEOUS	Granite.	

The Quantity of Excavation in Wells for Each Foot in Depth.

(Hurst.)

Diameter of Excavation.		Quantity.	Diameter of Excavation.		Quantity.
ft.	in.	cubic yards.	ft.	in.	cubic yards.
3	0	·2618	6	6	1·2290
3	3	·3072	6	9	1·3254
3	6	·3563	7	0	1·4254
3	9	·4091	7	3	1·5290
4	0	·4654	7	6	1·6362
4	3	·5254	7	9	1·7472
4	6	·5890	8	0	1·8617
4	9	·6563	8	6	2·1017
5	0	·7272	9	0	2·3562
5	3	·8018	9	6	2·6253
5	6	·8799	10	0	2·9089
5	9	·9617	10	6	3·2070
6	0	1·0472	11	0	3·5198
6	3	1·1363	12	0	4·1888

The Measure in Gallons, and the Weight in Pounds, of Water contained in Wells, for each Foot in Depth.

Diameter.		No. of Galls.	Weight.	Diameter.		No. of Galls.	Weight.
ft.	in.		lb.	ft.	in.		lb.
2	0	19·61	196·1	6	6	206·59	2065·9
2	6	30·56	305·6	7	0	239·05	2395·0
3	0	43·97	439·7	7	6	275·49	2754·9
3	6	60·00	600·0	8	0	313·43	3134·3
4	0	78·19	781·9	8	6	353·03	3533·0
4	6	98·87	988·7	9	0	395·42	3954·2
5	0	122·23	1222·3	9	6	441·71	4417·1
5	6	147·96	1479·6	10	0	489·93	4899·3
6	0	175·99	1759·9				

BRICKWORK.

THE NUMBER OF BRICKS AND QUANTITY OF BRICKWORK IN WELLS FOR EACH FOOT IN DEPTH.

(Hurst.)

	HALF-BRICK THICK.			ONE BRICK THICK.		
	Number of Bricks.		Cubic Feet of Brickwork.	Number of Bricks.		Cubic Feet of Brickwork.
	Laid Dry.	Laid in Mortar.		Laid Dry.	Laid in Mortar.	
1·0	28	23	1·6198	70	58	4·1233
1·3	33	27	1·8145	80	66	4·7124
1·6	38	31	2·2089	90	74	5·3015
1·9	43	35	2·5035	102	82	5·8905
2·0	48	41	2·7979	112	92	6·4795
2·3	53	44	3·0926	122	100	7·0686
2·6	58	48	3·3870	132	108	7·6577
3·0	68	57	3·9760	154	126	8·8357
3·6	79	65	4·5651	174	142	10·0139
4·0	89	73	5·1511	194	159	11·1919
4·6	100	82	5·7432	214	176	12·3704
5·0	110	90	6·3322	234	192	13·5481
5·6	120	98	6·9213	254	209	14·7263
6·0	130	107	7·5103	276	226	15·9043
6·6	140	115	8·0994	296	242	17·0825
7·0	150	123	8·6884	316	260	18·2605
7·6	160	131	9·2775	336	276	19·4387
8·0	170	140	9·8665	358	292	20·6167
8·6	180	148	10·4556	378	308	21·7949
9·0	191	156	11·0446	398	326	22·9729
10·0	212	174	12·2227	438	360	25·3291

Good bricks are characterised as being regular in shape, with plane parallel surfaces, and sharp right-angles; clear ringing sound when struck, a compact uniform structure when broken, and freedom from air-bubbles and cracks. They should not absorb more than one-fifteenth of their weight in water.

After making liberal allowance for waste, 9 bricks will build a square foot 9 inches thick, or 900, 100 square feet, or say 2880 to the rood of 9-inch work, which gives the simple rule of 80 bricks = a square yard of 9-inch work.

The resistance to crushing is from 1200 to 4500 lb. a square inch; the resistance to fracture, from 600 to 2500 lb. a square inch; tensile strength, 275 lb. a square inch; weight, in mortar, 175 lb. a cubic foot; in cement, 125 lb. a cubic foot.

Compressed bricks are much heavier, and consequently proportionately stronger, than those of ordinary make.

SUNDRY MEASURES OF WATER.

The weight of one gallon of water, at 62° F., is 10 pounds, and the correct volume is 277·123 cubic inches. The commonly accepted volume is 277·274 cubic inches.

One cubic foot of water contains 6·2355 gallons, or approximately 6¼ gallons.

The volume of water at 62° F. in cubic inches, multiplied by ·00036, gives the capacity in gallons.

The capacity of one gallon is equal to one square foot, about two inches deep; or to one circular foot about 2½ inches deep.

One ton of water, at 62° F., contains 224 gallons.

The volume of given weights of water, at 62·4 pounds a cubic foot are as follows ;—

1 ton, 35·90 cubic feet ; 1 cwt., 1·795 cubic feet ; 1 quarter, ·499 cubic feet; 1 pound, ·016 cubic foot, or 27·692 cubic inches.

36 cubic feet, or 1⅓ cubic yards, of water, at 62·4 pounds a cubic foot, weighs about one ton.

1 cubic yard of water weighs about 15 cwt., or ¾ ton. It is equal to 168·36 gallons.

1 cubic metre of water is equal in volume to 35·3156 cubic feet, or 220·09 gallons; and, at 62·4 pounds a cubic foot, it weighs one ton nearly (36 pounds less). It is nearly equivalent to the old English tun of 4 hogsheads, which is 210 imperial gallons, and is a better unit for measuring water-supply or sewage than the gallon.

A pipe one yard long holds about as many pounds of cold water as the square of its diameter in inches.

STORING WELL-WATER.

The reservoirs for storing well-water should be covered with brick arches, as the water is generally found to become rapidly impure on being exposed to the sunlight, principally owing to the rapid growth of vegetation. Various methods have been tried, such as keeping up a constant current of fresh water through them, and a liberal use of caustic lime; but so rapid is the growth of the vegetation, as well as the change in the colour of the water, that a few hours of bright sunlight may suffice to spoil several million gallons. These bad results are completely prevented by covering the reservoirs.

HINTS ON SUPERINTENDING WELL-WORK.

The engineer who has to superintend the construction of a well should be ever on the watch to see whether, in the course of the work, the strata become so modified as to overthrow conclusions previously arrived at, and on account of which the well has been undertaken.

A journal of everything connected with the work should be carefully made, and if this one point alone is attended to, it will be found of great service both for present and future reference.

Before commencing a well a wooden box should be provided, divided by a number of partitions into small boxes; these serve to keep specimens of the strata, which should be numbered consecutively, and described against corresponding numbers in the journal. At each change of character in the strata, as well as every time the boring rods are drawn to surface, the soil should be carefully examined, and at each change a small quantity placed in one of the divisions of the core box, noting the depth at which it was obtained, with other necessary particulars. A note should be made of all the different water levels passed through, the height of the well above the river near which it is situated, as well as its height above the sea. The memoranda in the journal relating to accidents should be especially clear and distinct in their details; it is necessary to describe the effects of each tool used in the search for, or recovery of, broken tools in a borehole, in order to suit the case with the proper appliances, for without precaution we may seek for a tool indefinitely without being sure of touching it, and perhaps aggravate the evil instead of remedying it. It is by no means a bad plan to make rough notes of all immediate remarks or impressions, in such a manner as to form a full and detailed account of any incidents which occur either in raising or lowering the tools. At the time of an accident a well-kept journal is a precious resource, and at a given moment all previous observations, trivial as they may have often seemed, will form a valuable clue to explain difficulties, without this aid perfectly inexplicable.

When an engineer has a certain latitude allowed him in the choice of a position for a well, he should not, other things being equal, neglect the advantages which will be derived from the proximity of a road for the transport of his supplies ; of a well, if not a brook, from which to obtain the water necessary for the cleansing of the tools ; and of a neighbouring dwelling, to facilitate his active supervision. This supervision, having often to be carried on both day and night, should be the object of particular study ; well carried out, it may be effective, while at the same time allowing a great amount of liberty ; badly carried out, however fatiguing it may be, it will be incomplete.

RATE OF PROGRESS OF BORING.
(André.)

There are probably no engineering operations in which the rate of progress is so variable as it is in that of boring. That such must necessarily be the case will be obvious when we bear in mind that the strata composing the earth's crust consist of very different materials; that these materials are mingled in very different proportions, and that they have in different parts been subjected to the action of very different agencies operating with very different degrees of intensity. Hence it arises not only that some kinds of rocks require a much longer time to bore through than others, but also that the length of time may

s

vary in rocks of the same character, and that the character may change within a short horizontal distance. Thus it is utterly impossible to predicate concerning the length of time which a boring in an unknown district may occupy, and only a rough approximation can be arrived at in the case of localities whose geological constitution has been generally determined. Such an approximation may, however, be attained to, and it is useful in estimating the probable cost; and to attain the same end, for unknown localities, an average may be taken of the time required in districts of a similar geological character. The following, which are given for this purpose, are the averages of a great number of borings executed under various conditions by the ordinary methods. The progress indicated represents that made in one day of eleven hours.

				ft.	in.
1. Tertiary and Cretaceous Strata, to a depth of 100 yards, average progress				1	8
2. Cretaceous Strata, without flints	,,	250	,, ,,	2	1
3. Cretaceous Strata, with flints	,,	250	,, ,,	1	4
4. New Red Sandstone	,,	250	,, ,,	1	10
5. New Red Sandstone	,,	500	,, ,,	1	5
6. Permian Strata	,,	250	,, ,,	2	0
7. Coal Measures	,,	200	,, ,,	2	3
8. Coal Measures	,,	400	,, ,,	1	8
General Average ..		275		1	9

When the cost of materials and labour is known, that of the boring may be approximately estimated from the above averages. Should hard limestone or igneous rock be met with, the rate of progress may be less than half the above general average. Below 100 yards, not only does the rate of progress rapidly increase, but the material required diminishes in like proportion, so that for superficial borings no surface erections are needed, and the cost sinks to two or three shillings a yard.

COST OF BORING.

The cost of boring when executed by contract has already been treated of at page 94. The following formula will furnish the same results as the rule there given, but with the least possible labour of calculation;

$$x = 0{\cdot}5\,d\,({\cdot}187 + {\cdot}0187\,d);$$

x being the sum sought, in pounds, and d the depth of the boring in yards.

Example.   Let it be required to know the cost of a borehole 250 yards deep.

Here $125 \{ \cdot 187 + (\cdot 0187 \times 250) \} = £607 \cdot 75.$

TEMPERING BORING CHISELS.

1. Heat the chisel to a blood-red heat, and then hammer it until nearly cold; again, heat it to a blood red and quench as quickly as possible in 3 gallons of water in which is dissolved 2 oz. of oil of vitriol, 2 oz. of soda, and ½ oz. of saltpetre, or 2 oz. of sal ammoniac, 2 oz. of spirit of nitre, 1 oz. of oil of vitriol: the chisel to remain in the liquor until it is cold.

2. To 3 gallons of water add 3 oz. of spirit of nitre, 3 oz. of spirits of hartshorn, 3 oz. of white vitriol, 3 oz. of sal ammoniac, 3 oz. of alum, 6 oz. of salt, with a double handful of hoof-parings, the chisel to be heated to a dark cherry red.

GASES IN WELLS.

The most abundant deleterious gas met with in wells is carbonic acid, which extinguishes flame and is fatal to animal life. Carbonic acid is most frequently met with in the chalk, where it has been found to exist in greater quantity in the lower than in the upper portion of the formation, and in that division to be unequally distributed.   Fatal effects from it at Epsom, 200 feet down, and in Norbury Park, near Dorking, 400 feet down, have been recorded.   At Bexley Heath, after sinking through 140 feet of gravel and sand and 30 feet of chalk, it rushed out and extinguished the candles of the workmen.   Air mixed with one-tenth of this gas will extinguish lights; it is very poisonous, and when the atmosphere contains 8 per cent. or more there is danger of suffocation.   When present it is found most abundantly in the lower parts of a well from its great specific gravity.

Sulphuretted hydrogen is also occasionally met with, and is supposed to be generated from the decomposition of water and iron pyrites.

In districts in which the chalk is covered with sand and London clay, carburetted hydrogen is occasionally emitted, but

s 2

more frequently sulphuretted hydrogen. Carburetted hydrogen seldom inflames in wells, but in making the Thames Tunnel it sometimes issued in such abundance as to explode by the lights and scorch the workmen. Sulphuretted hydrogen also streamed out in the same place, but in no instance with fatal effects. At Ash, near Farnham, a well was dug in sand to the depth of 36 feet, and one of the workmen descending into it was instantly suffocated. Fatal effects have also resulted elsewhere from the accumulation of this gas in wells.

SPECIFICATION AND TENDER.

The following form of specification and tender is one which has been frequently employed by the writer. In this particular instance it is filled in for work in cretaceous strata, the modifications necessary for application to another case are sufficiently obvious :—

SPECIFICATION TO BE OBSERVED BY THE CONTRACTOR IN SINKING AND BORING A WELL ON THE ESTATE OF ——, SITUATE IN THE PARISH OF ——, IN THE COUNTY OF ——.

Position. The Well is to be sunk and bored at the spot W, coloured red upon the plan to be furnished to the contracting parties.

Nature of strata to be passed through. The Strata to be passed through consist of about 6 feet of Tertiary deposits, the Upper and the Lower Chalk, and the underlying Upper Greensand.

Depth. The Well is to be sunk in the Chalk to a depth of 260 feet, and from this depth the well is to be carried down by boring to the bottom of the Upper Greensand. The estimated depth of the boring is 70 feet.

Dimensions. The Well is to have a clear diameter of 6 feet, and to be lined with bricks, 9-inch work, well laid in cement, to a depth of 20 feet from surface. The borehole is to have a diameter of not less than 4 inches, and to be tubed throughout with iron tubing.

General conditions. The Contractor will be required to find all labour, tools, appliances, and apparatus or materials of whatever kind or description, required for the due and full performance of his contract, together with any transport or carriage in connection therewith ; he will further be required to restore the surface of the land, and make good any damage in reference thereto, as well as to restore any fences or any damage of whatever nature that may be caused in connection with the work.

Should the Contractor, after three days' notice in writing under the hand of the Engineer, fail to carry out any of the provisions of this Specification, the Engineer may take charge of and proceed with the work at the cost of the Contractor, and for that purpose may take and use without hindrance any tools, appliances, apparatus, or materials, upon the works belonging to the Contractor.

Should a sufficient quantity of water be met with short of the Greensand the Engineer reserves to himself the right of stopping the boring at any point.

The Contractor, on the acceptance of his Tender, will be Time. required to proceed with the work forthwith, and to complete the whole of the work within twelve weeks from the date of the acceptance of his Tender.

Payments on account will be made on the certificate of Paymen the Engineer, after the first 100 feet have been sunk, to the extent of 75 per cent. of the accepted price, the balance to be paid on the completion of the work to the satisfaction of the Engineer.

Persons tendering are to state the price of the sinking for Tenders. the whole depth of 260 feet, and the price of the boring for each 20 feet in depth.

Tender.

—— hereby undertake and agree to sink and bore a Well in the situation and of the depth required, and provide all superintendence, labour, tools, appliances, apparatus, materials, and carriage in connection with the work, and in accordance with the full terms and conditions of the annexed Specification, at the several rates or prices respectively stated in the Schedules numbered 1 and 2; and ——— hereby further undertake and agree to execute the work in the time stated in such Specification to the satisfaction of the Engineer appointed to superintend the same.

Schedules referred to in ——— Tender of_____

SCHEDULE No. 1.

Description of Work.	Price.		
Sinking shaft 6 feet diameter, 260 feet, 20-feet steining, 9-inch brickwork set in cement }			

SCHEDULE No. 2.

Description of Work.	Price.		
Boring for each 20 feet depth, and lining with cast-iron tubes.. .. .. }			

INDEX.

266 INDEX.

LONDON: PRINTED BY WILLIAM CLOWES AND SONS, LIMITED,
STAMFORD STREET AND CHARING CROSS.

BOOKS RELATING

TO

APPLIED SCIENCE

PUBLISHED BY

E. & F. N. SPON,

LONDON: 125, STRAND.

NEW YORK: 35, MURRAY STREET.

———•———

A Pocket-Book for Chemists, Chemical Manufacturers,
Metallurgists, Dyers, Distillers, Brewers, Sugar Refiners, Photographers,
Students, etc., etc. By THOMAS BAYLEY, Assoc. R.C. Sc. Ireland, Ana-
lytical and Consulting Chemist and Assayer.   Third edition, with
additions, 437 pp., royal 32mo, roan, gilt edges, 5*s.*

SYNOPSIS OF CONTENTS :

Atomic Weights and Factors—Useful Data—Chemical Calculations—Rules for Indirect
Analysis—Weights and Measures—Thermometers and Barometers—Chemical Physics—
Boiling Points, etc.—Solubility of Substances—Methods of Obtaining Specific Gravity—Con-
version of Hydrometers—Strength of Solutions by Specific Gravity—Analysis—Gas Analysis—
Water Analysis—Qualitative Analysis and Reactions—Volumetric Analysis—Manipulation—
Mineralogy — Assaying — Alcohol — Beer — Sugar — Miscellaneous Technological matter
relating to Potash, Soda, Sulphuric Acid, Chlorine, Tar Products, Petroleum, Milk, Tallow,
Photography, Prices, Wages, Appendix, etc., etc.

The Mechanician : A Treatise on the Construction
and Manipulation of Tools, for the use and instruction of Young Engineers
and Scientific Amateurs, comprising the Arts of Blacksmithing and Forging ; the Construction and Manufacture of Hand Tools, and the various
Methods of Using and Grinding them ; the Construction of Machine Tools,
and how to work them ; Machine Fitting and Erection ; description of
Hand and Machine Processes ; Turning and Screw Cutting ; principles of
Constructing and details of Making and Erecting Steam Engines, and the
various details of setting out work, etc., etc. By CAMERON KNIGHT,
Engineer. *Containing* 1147 *illustrations*, and 397 pages of letter-press.
Third edition, 4to, cloth, 18*s.*

B

On Designing Belt Gearing. By E. J. COWLING WELCH, Mem. Inst. Mech. Engineers, Author of 'Designing Valve Gearing.' Fcap. 8vo, sewed, 6d.

A Handbook of Formulæ, Tables, and Memoranda, *for Architectural Surveyors and others engaged in Building.* By J. T. HURST, C.E. Thirteenth edition, royal 32mo, roan, 5s.

"It is no disparagement to the many excellent publications we refer to, to say that in our opinion this little pocket-book of Hurst's is the very best of them all, without any exception. It would be useless to attempt a recapitulation of the contents, for it appears to contain almost *everything* that anyone connected with building could require, and, best of all, made up in a compact form for carrying in the pocket, measuring only 5 in. by 3 in., and about ¼ in. thick, in a limp cover. We congratulate the author on the success of his laborious and practically compiled little book, which has received unqualified and deserved praise from every professional person to whom we have shown it."—*The Dublin Builder.*

Tabulated Weights of Angle, Tee, Bulb, Round, *Square, and Flat Iron and Steel,* and other information for the use of Naval Architects and Shipbuilders. By C. H. JORDAN, M.I.N.A. Fourth edition, 32mo, cloth, 2s. 6d.

Quantity Surveying. By J. LEANING. With 42 illustrations, crown 8vo, cloth, 9s.

CONTENTS :

A complete Explanation of the London Practice.
General Instructions.
Order of Taking Off.
Modes of Measurement of the various Trades.
Use and Waste.
Ventilation and Warming.
Credits, with various Examples of Treatment.
Abbreviations.
Squaring the Dimensions.
Abstracting, with Examples in illustration of each Trade.
Billing.
Examples of Preambles to each Trade.
Form for a Bill of Quantities.
Do.   Bill of Credits.
Do.   Bill for Alternative Estimate.
Restorations and Repairs, and Form of Bill.
Variations before Acceptance of Tender.
Errors in a Builder's Estimate.

Schedule of Prices.
Form of Schedule of Prices.
Analysis of Schedule of Prices.
Adjustment of Accounts.
Form of a Bill of Variations.
Remarks on Specifications.
Prices and Valuation of Work, with Examples and Remarks upon each Trade.
The Law as it affects Quantity Surveyors, with Law Reports.
Taking Off after the Old Method.
Northern Practice.
The General Statement of the Method recommended by the Manchester Society of Architects for taking Quantities.
Examples of Collections.
Examples of "Taking Off" in each Trade.
Remarks on the Past and Present Method of Estimating.

A Practical Treatise on Heat, as applied to the *Useful Arts;* for the Use of Engineers, Architects, &c. By THOMAS BOX. *With* 14 *plates.* Third edition, crown 8vo, cloth, 12s. 6d.

A Descriptive Treatise on Mathematical Drawing *Instruments:* their construction, uses, qualities, selection, preservation, and suggestions for improvements, with hints upon Drawing and Colouring. By W. F. STANLEY, M.R.I. Fifth edition, *with numerous illustrations,* crown 8vo, cloth, 5s.

Spons' Architects' and Builders' Pocket-Book of Prices and Memoranda. Edited by W. YOUNG, Architect. Royal 32mo, roan, 4s. 6d. ; or cloth, red edges, 3s. 6d. *Published annually.* Eleventh edition. *Now ready.*

Long-Span Railway Bridges, comprising Investigations of the Comparative Theoretical and Practical Advantages of the various adopted or proposed Type Systems of Construction, with numerous Formulæ and Tables giving the weight of Iron or Steel required in Bridges from 300 feet to the limiting Spans ; to which are added similar Investigations and Tables relating to Short-span Railway Bridges. Second and revised edition. By B. BAKER, Assoc. Inst. C.E. *Plates,* crown 8vo, cloth, 5s.

Elementary Theory and Calculation of Iron Bridges and Roofs. By AUGUST RITTER, Ph.D., Professor at the Polytechnic School at Aix-la-Chapelle. Translated from the third German edition, by H. R. SANKEY, Capt. R.E. With 500 *illustrations,* 8vo, cloth, 15s.

The Builder's Clerk : a Guide to the Management of a Builder's Business. By THOMAS BALES. Fcap. 8vo, cloth, 1s. 6d.

The Elementary Principles of Carpentry. By THOMAS TREDGOLD. Revised from the original edition, and partly re-written, by JOHN THOMAS HURST. Contained in 517 pages of letter-press, and *illustrated with 48 plates and 150 wood engravings.* Third edition, crown 8vo, cloth, 18s.

Section I. On the Equality and Distribution of Forces—Section II. Resistance of Timber—Section III. Construction of Floors—Section IV. Construction of Roofs—Section V. Construction of Domes and Cupolas—Section VI. Construction of Partitions—Section VII. Scaffolds, Staging, and Gantries—Section VIII. Construction of Centres for Bridges—Section IX. Coffer-dams, Shoring, and Strutting—Section X. Wooden Bridges and Viaducts—Section XI. Joints, Straps, and other Fastenings—Section XII. Timber.

Our Factories, Workshops, and Warehouses : their Sanitary and Fire-Resisting Arrangements. By B. H. THWAITE, Assoc. Mem. Inst. C.E. *With 183 wood engravings,* crown 8vo, cloth, 9s.

Gold : Its Occurrence and Extraction, embracing the Geographical and Geological Distribution and the Mineralogical Characters of Gold-bearing rocks ; the peculiar features and modes of working Shallow Placers, Rivers, and Deep Leads ; Hydraulicing ; the Reduction and Separation of Auriferous Quartz ; the treatment of complex Auriferous ores containing other metals ; a Bibliography of the subject and a Glossary of Technical and Foreign Terms. By ALFRED G. LOCK, F.R.G.S. *With numerous illustrations and maps,* 1250 pp., super-royal 8vo, cloth, 2l. 12s. 6d.

A Practical Treatise on Coal Mining. By GEORGE
G. ANDRÉ, F.G.S., Assoc. Inst. C.E., Member of the Society of Engineers.
With 82 lithographic plates. 2 vols., royal 4to, cloth, 3*l.* 12*s.*

Iron Roofs : Examples of Design, Description. *Illustrated with 64 Working Drawings of Executed Roofs.* By ARTHUR T.
WALMISLEY, Assoc. Mem. Inst. C.E. Imp. 4to, half-morocco, £2 12*s.* 6*d.*

A History of Electric Telegraphy, to the Year 1837.
Chiefly compiled from Original Sources, and hitherto Unpublished Docu-
ments, by J. J. FAHIE, Mem. Soc. of Tel. Engineers, and of the Inter-
national Society of Electricians, Paris. Crown 8vo, cloth, 9*s.*

Spons' Information for Colonial Engineers. Edited
by J. T. HURST. Demy 8vo, sewed.

No. 1, Ceylon. By ABRAHAM DEANE, C.E. 2*s.* 6*d.*

CONTENTS :

Introductory Remarks—Natural Productions—Architecture and Engineering—Topo-
graphy, Trade, and Natural History—Principal Stations—Weights and Measures, etc., etc.

No. 2. Southern Africa, including the Cape Colony, Natal, and the
Dutch Republics. By HENRY HALL, F.R.G.S., F.R.C.I. With
Map. 3*s.* 6*d.*

CONTENTS :

General Description of South Africa—Physical Geography with reference to Engineering
Operations—Notes on Labour and Material in Cape Colony—Geological Notes on Rock
Formation in South Africa—Engineering Instruments for Use in South Africa—Principal
Public Works in Cape Colony : Railways, Mountain Roads and Passes, Harbour Works,
Bridges, Gas Works, Irrigation and Water Supply, Lighthouses, Drainage and Sanitary
Engineering, Public Buildings, Mines—Table of Woods in South Africa—Animals used for
Draught Purposes—Statistical Notes—Table of Distances—Rates of Carriage, etc.

No. 3. India. By F. C. DANVERS, Assoc. Inst. C.E. With Map. 4*s.* 6*d.*

CONTENTS :

Physical Geography of India—Building Materials—Roads—Railways—Bridges—Irrigation—River Works—Harbours—Lighthouse Buildings—Native Labour—The Principal
Trees of India—Money—Weights and Measures—Glossary of Indian Terms, etc.

A Practical Treatise on Casting and Founding,
including descriptions of the modern machinery employed in the art. By
N. E. SPRETSON, Engineer. Third edition, with 82 *plates* drawn to
scale, 412 pp., demy 8vo, cloth, 18*s.*

Steam Heating for Buildings ; or, Hints to Steam
Fitters, being a description of Steam Heating Apparatus for Warming
and Ventilating Private Houses and Large Buildings, with remarks on
Steam, Water, and Air in their relation to Heating. By W. J. BALDWIN.
With many illustrations. Fourth edition, crown 8vo, cloth, 10*s.* 6*d.*

The Depreciation of Factories and their Valuation.
By EWING MATHESON, M. Inst. C.E. 8vo, cloth, 6s.

A Handbook of Electrical Testing. By H. R. KEMPE,
M.S.T.E. Third edition, revised and enlarged, crown 8vo, cloth, 15s.

Gas Works: their Arrangement, Construction, Plant,
and Machinery. By F. COLYER, M. Inst. C.E. *With 31 folding plates,*
8vo, cloth, 24s.

The Clerk of Works: a Vade-Mecum for all engaged
in the Superintendence of Building Operations. By G. G. HOSKINS,
F.R.I.B.A. Third edition, fcap. 8vo, cloth, 1s. 6d.

American Foundry Practice: Treating of Loam,
Dry Sand, and Green Sand Moulding, and containing a Practical Treatise
upon the Management of Cupolas, and the Melting of Iron. By T. D.
WEST, Practical Iron Moulder and Foundry Foreman. Second edition,
with numerous illustrations, crown 8vo, cloth, 10s. 6d.

The Maintenance of Macadamised Roads. By T.
CODRINGTON, M.I.C.E, F.G.S., General Superintendent of County Roads
for South Wales. 8vo, cloth, 6s.

Hydraulic Steam and Hand Power Lifting and
Pressing Machinery. By FREDERICK COLYER, M. Inst. C.E., M. Inst. M.E.
With 73 plates, 8vo, cloth, 18s.

Pumps and Pumping Machinery. By F. COLYER,
M.I.C.E., M.I.M.E. *With 23 folding plates,* 8vo, cloth, 12s. 6d.

The Municipal and Sanitary Engineer's Handbook.
By H. PERCY BOULNOIS, Mem. Inst. C.E., Borough Engineer, Ports-
mouth. *With numerous illustrations,* demy 8vo, cloth, 12s. 6d.

CONTENTS:

The Appointment and Duties of the Town Surveyor—Traffic—Macadamised Roadways—
Steam Rolling—Road Metal and Breaking—Pitched Pavements—Asphalte—Wood Pavements
—Footpaths—Kerbs and Gutters—Street Naming and Numbering—Street Lighting—Sewer-
age—Ventilation of Sewers—Disposal of Sewage—House Drainage—Disinfection—Gas and
Water Companies, &c., Breaking up Streets—Improvement of Private Streets—Borrowing
Powers—Artizans' and Labourers' Dwelling—Public Conveniences—Scavenging, including
Street Cleansing—Watering and the Removing of Snow—Planting Street Trees—Deposit of
Plans—Dangerous Buildings—Hoardings—Obstructions—Improving Street Lines—Cellar
Openings—Public Pleasure Grounds—Cemeteries—Mortuaries—Cattle and Ordinary Markets
—Public Slaughter-houses, etc.—Giving numerous Forms of Notices, Specifications, and
General Information upon these and other subjects of great importance to Municipal Engi-
neers and others engaged in Sanitary Work.

Tables of the Principal Speeds occurring in Mechanical Engineering, expressed in metres in a second. By P. KEERAYEFF, Chief Mechanic of the Obouchoff Steel Works, St. Petersburg ; translated by SERGIUS KERN, M.E. Fcap. 8vo, sewed, 6*d.*

A Treatise on the Origin, Progress, Prevention, and Cure of Dry Rot in Timber; with Remarks on the Means of Preserving Wood from Destruction by Sea-Worms, Beetles, Ants, etc. By THOMAS ALLEN BRITTON, late Surveyor to the Metropolitan Board of Works, etc., etc. *With* 10 *plates,* crown 8vo, cloth, 7*s.* 6*d.*

Metrical Tables. By G. L. MOLESWORTH, M.I.C.E. 32mo, cloth, 1*s.* 6*d.*

CONTENTS.

General—Linear Measures—Square Measures—Cubic Measures—Measures of Capacity—Weights—Combinations—Thermometers.

Elements of Construction for Electro-Magnets. By Count TH. DU MONCEL, Mem. de l'Institut de France. Translated from the French by C. J. WHARTON. Crown 8vo, cloth, 4*s.* 6*d.*

Electro-Telegraphy. By FREDERICK S. BEECHEY, Telegraph Engineer. A Book for Beginners. *Illustrated.* Fcap. 8vo, sewed, 6*d.*

Handrailing: by the Square Cut. By JOHN JONES, Staircase Builder. Fourth edition, *with seven plates,* 8vo, cloth, 3*s.* 6*d.*

Handrailing: by the Square Cut. By JOHN JONES, Staircase Builder. Part Second, *with eight plates,* 8vo, cloth, 3*s.* 6*d.*

Practical Electrical Units Popularly Explained, with *numerous illustrations* and Remarks. By JAMES SWINBURNE, late of J. W. Swan and Co., Paris, late of Brush-Swan Electric Light Company, U.S.A. 18mo, cloth, 1*s.* 6*d.*

Philipp Reis, Inventor of the Telephone: A Biographical Sketch. With Documentary Testimony, Translations of the Original Papers of the Inventor, &c. By SILVANUS P. THOMPSON, B.A., Dr. Sc., Professor of Experimental Physics in University College, Bristol. *With illustrations,* 8vo, cloth, 7*s.* 6*d.*

A Treatise on the Use of Belting for the Transmission of Power. By J. H. COOPER. Second edition, *illustrated,* 8vo, cloth, 15*s.*

A Pocket-Book of Useful Formulæ and Memoranda

for Civil and Mechanical Engineers. By GUILFORD L. MOLESWORTH, Mem. Inst. C.E., Consulting Engineer to the Government of India for State Railways. *With numerous illustrations*, 744 pp., Twenty-first edition, revised and enlarged, 32mo, roan, 6s.

SYNOPSIS OF CONTENTS:

Surveying, Levelling, etc.—Strength and Weight of Materials—Earthwork, Brickwork, Masonry, Arches, etc.—Struts, Columns, Beams, and Trusses—Flooring, Roofing, and Roof Trusses—Girders, Bridges, etc.—Railways and Roads—Hydraulic Formulæ—Canals, Sewers, Waterworks, Docks—Irrigation and Breakwaters—Gas, Ventilation, and Warming—Heat, Light, Colour, and Sound—Gravity : Centres, Forces, and Powers—Millwork, Teeth of Wheels, Shafting, etc.—Workshop Recipes—Sundry Machinery—Animal Power—Steam and the Steam Engine—Water-power, Water-wheels, Turbines, etc.—Wind and Windmills—Steam Navigation, Ship Building, Tonnage, etc.—Gunnery, Projectiles, etc.—Weights, Measures, and Money—Trigonometry, Conic Sections, and Curves—Telegraphy—Mensuration—Tables of Areas and Circumference, and Arcs of Circles—Logarithms, Square and Cube Roots, Powers—Reciprocals, etc.—Useful Numbers—Differential and Integral Calculus—Algebraic Signs—Telegraphic Construction and Formulæ.

Spons' Tables and Memoranda for Engineers ;

selected and arranged by J. T. HURST, C.E., Author of 'Architectural Surveyors' Handbook,' 'Hurst's Tredgold's Carpentry,' etc. Fifth edition, 64mo, roan, gilt edges, 1s. ; or in cloth case, 1s. 6d.

This work is printed in a pearl type, and is so small, measuring only 2½ in. by 1½ in. by ½ in. thick, that it may be easily carried in the waistcoat pocket.

"It is certainly an extremely rare thing for a reviewer to be called upon to notice a volume measuring but 2½ in. by 1½ in., yet these dimensions faithfully represent the size of the handy little book before us. The volume—which contains 118 printed pages, besides a few blank pages for memoranda—is, in fact, a true pocket-book, adapted for being carried in the waistcoat pocket, and containing a far greater amount and variety of information than most people would imagine could be compressed into so small a space. . . . . The little volume has been compiled with considerable care and judgment, and we can cordially recommend it to our readers as a useful little pocket companion."—*Engineering.*

A Practical Treatise on Natural and Artificial

Concrete, its Varieties and Constructive Adaptations. By HENRY REID, Author of the 'Science and Art of the Manufacture of Portland Cement.' New Edition, *with 59 woodcuts and 5 plates*, 8vo, cloth, 15s.

Hydrodynamics : Treatise relative to the Testing of

Water-Wheels and Machinery, with various other matters pertaining to Hydrodynamics. By JAMES EMERSON. *With numerous illustrations*, 360 pp. Third edition, crown 8vo, cloth, 4s. 6d.

Electricity as a Motive Power. By Count TH. DU

MONCEL, Membre de l'Institut de France, and FRANK GERALDY, Ingénieur des Ponts et Chaussées. Translated and Edited, with Additions, by C. J. WHARTON, Assoc. Soc. Tel. Eng. and Elec. *With 113 engravings and diagrams*, crown 8vo, cloth, 7s. 6d.

Hints on Architectural Draughtsmanship. By G. W.

TUXFORD HALLATT. Fcap. 8vo, cloth, 1s. 6d.

Treatise on Valve-Gears, with special consideration of the Link-Motions of Locomotive Engines. By Dr. GUSTAV ZEUNER, Professor of Applied Mechanics at the Confederated Polytechnikum of Zurich. Translated from the Fourth German Edition, by Professor J. F. KLEIN, Lehigh University, Bethlehem, Pa. *Illustrated*, 8vo, cloth, 12s. 6d.

The French-Polisher's Manual. By a French-Polisher; containing Timber Staining, Washing, Matching, Improving, Painting, Imitations, Directions for Staining, Sizing, Embodying, Smoothing, Spirit Varnishing, French-Polishing, Directions for Re-polishing. Third edition, royal 32mo, sewed, 6d.

Hops, their Cultivation, Commerce, and Uses in various Countries. By P. L. SIMMONDS. Crown 8vo, cloth, 4s. 6d.

A Practical Treatise on the Manufacture and Distri-bution of Coal Gas. By WILLIAM RICHARDS. Demy 4to, with *numerous wood engravings and 29 plates,* cloth, 28s.

SYNOPSIS OF CONTENTS:

Introduction—History of Gas Lighting—Chemistry of Gas Manufacture, by Lewis Thompson, Esq., M.R.C.S.—Coal, with Analyses, by J. Paterson, Lewis Thompson, and G. R. Hislop, Esqrs.—Retorts, Iron and Clay—Retort Setting—Hydraulic Main—Con-densers—Exhausters—Washers and Scrubbers—Purifiers—Purification—History of Gas Holder—Tanks, Brick and Stone, Composite, Concrete, Cast-iron, Compound Annular Wrought-iron—Specifications—Gas Holders—Station Meter—Governor—Distribution—Mains—Gas Mathematics, or Formulæ for the Distribution of Gas, by Lewis Thompson, Esq.—Services—Consumers' Meters—Regulators—Burners—Fittings—Photometer—Carburization of Gas—Air Gas and Water Gas—Composition of Coal Gas, by Lewis Thompson, Esq.—Analyses of Gas—Influence of Atmospheric Pressure and Temperature on Gas—Residual Products—Appendix—Description of Retort Settings, Buildings, etc., etc.

Practical Geometry, Perspective, and Engineering Drawing; a Course of Descriptive Geometry adapted to the Require-ments of the Engineering Draughtsman, including the determination of cast shadows and Isometric Projection, each chapter being followed by numerous examples; to which are added rules for Shading, Shade-lining, etc., together with practical instructions as to the Lining, Colouring, Printing, and general treatment of Engineering Drawings, with a chapter on drawing Instruments. By GEORGE S. CLARKE, Capt. R.E. Second edition, *with 21 plates.* 2 vols., cloth, 10s. 6d.

The Elements of Graphic Statics. By Professor KARL VON OTT, translated from the German by G. S. CLARKE, Capt. R.E., Instructor in Mechanical Drawing, Royal Indian Engineering College. *With 93 illustrations,* crown 8vo, cloth, 5s.

The Principles of Graphic Statics. By GEORGE SYDENHAM CLARKE, Capt. Royal Engineers. *With 112 illustrations.* 4to, cloth, 12s. 6d.

Dynamo-Electric Machinery: A Manual for Students of Electro-technics. By SILVANUS P. THOMPSON, B.A., D.Sc., Professor of Experimental Physics in University College, Bristol, etc., etc. *Illus-trated,* 8vo, cloth, 12s. 6d.

The New Formula for Mean Velocity of Discharge of Rivers and Canals. By W. R. KUTTER. Translated from articles in the 'Cultur-Ingénieur,' by LOWIS D'A. JACKSON, Assoc. Inst. C.E. 8vo, cloth, 12s. 6d.

Practical Hydraulics; a Series of Rules and Tables for the use of Engineers, etc., etc. By THOMAS BOX. Fifth edition, *numerous plates*, post 8vo, cloth, 5s.

A Practical Treatise on the Construction of Horizontal and Vertical Waterwheels, specially designed for the use of operative mechanics. By WILLIAM CULLEN, Millwright and Engineer. *With* 11 *plates*. Second edition, revised and enlarged, small 4to, cloth, 12s. 6d.

Tin: Describing the Chief Methods of Mining, Dressing and Smelting it abroad ; with Notes upon Arsenic, Bismuth and Wolfram. By ARTHUR G. CHARLETON, Mem. American Inst. of Mining Engineers. *With plates*, 8vo, cloth, 12s. 6d.

Perspective, Explained and Illustrated. By G. S. CLARKE, Capt. R.E. *With illustrations*, 8vo, cloth, 3s. 6d.

The Essential Elements of Practical Mechanics; *based on the Principle of Work*, designed for Engineering Students. By OLIVER BYRNE, formerly Professor of Mathematics, College for Civil Engineers. Third edition, *with* 148 *wood engravings*, post 8vo, cloth, 7s. 6d.

CONTENTS :

Chap. 1. How Work is Measured by a Unit, both with and without reference to a Unit of Time—Chap. 2. The Work of Living Agents, the Influence of Friction, and introduces one of the most beautiful Laws of Motion—Chap. 3. The principles expounded in the first and second chapters are applied to the Motion of Bodies—Chap. 4. The Transmission of Work by simple Machines—Chap. 5. Useful Propositions and Rules.

The Practical Millwright and Engineer's Ready Reckoner; or Tables for finding the diameter and power of cog-wheels, diameter, weight, and power of shafts, diameter and strength of bolts, etc. By THOMAS DIXON. Fourth edition, 12mo, cloth, 3s.

Breweries and Maltings : their Arrangement, Construction, Machinery, and Plant. By G. SCAMELL, F.R.I.B.A. Second edition, revised, enlarged, and partly rewritten. By F. COLYER, M.I.C.E., M.I.M.E. *With* 20 *plates*, 8vo, cloth, 18s.

A Practical Treatise on the Manufacture of Starch, *Glucose, Starch-Sugar, and Dextrine*, based on the German of L. Von Wagner, Professor in the Royal Technical School, Buda Pesth, and other authorities. By JULIUS FRANKEL ; edited by ROBERT HUTTER, proprietor of the Philadelphia Starch Works. *With* 58 *illustrations*, 344 pp., 8vo, cloth, 18s.

A Practical Treatise on Mill-gearing, Wheels, Shafts,
Riggers, etc.; for the use of Engineers. By THOMAS BOX. Third
edition, *with 11 plates.* Crown 8vo, cloth, 7s. 6d.

Mining Machinery: a Descriptive Treatise on the
Machinery, Tools, and other Appliances used in Mining. By G. G.
ANDRÉ, F.G.S., Assoc. Inst. C.E., Mem. of the Society of Engineers.
Royal 4to, uniform with the Author's Treatise on Coal Mining, con-
taining 182 *plates,* accurately drawn to scale, with descriptive text, in
2 vols., cloth, 3l. 12s.

CONTENTS :

Machinery for Prospecting, Excavating, Hauling, and Hoisting—Ventilation—Pumping—
Treatment of Mineral Products, including Gold and Silver, Copper, Tin, and Lead, Iron.
Coal, Sulphur, China Clay, Brick Earth, etc.

Tables for Setting out Curves for Railways, Canals,
Roads, etc., varying from a radius of five chains to three miles. By A.
KENNEDY and R. W. HACKWOOD. *Illustrated,* 32mo, cloth, 2s. 6d.

The Science and Art of the Manufacture of Portland
Cement, with observations on some of its constructive applications. *With*
66 *illustrations.* By HENRY REID, C.E., Author of 'A Practical
Treatise on Concrete,' etc., etc. 8vo, cloth, 18s.

The Draughtsman's Handbook of Plan and Map
Drawing; including instructions for the preparation of Engineering,
Architectural, and Mechanical Drawings. *With numerous illustrations*
in the text, and 33 *plates* (15 *printed in colours*). By G. G. ANDRÉ,
F.G.S., Assoc. Inst. C.E. 4to, cloth, 9s.

CONTENTS :

The Drawing Office and its Furnishings—Geometrical Problems—Lines, Dots, and their
Combination—Colours, Shading, Lettering, Bordering, and North Points—Scales—Plotting
—Civil Engineers' and Surveyors' Plans—Map Drawing—Mechanical and Architectural
Drawing—Copying and Reducing Trigonometrical Formulæ, etc., etc.

The Boiler-maker's and Iron Ship-builder's Companion,
comprising a series of original and carefully calculated tables, of the
utmost utility to persons interested in the iron trades. By JAMES FODEN,
author of 'Mechanical Tables,' etc. Second edition revised, *with illustra-*
tions, crown 8vo, cloth, 5s.

Rock Blasting: a Practical Treatise on the means
employed in Blasting Rocks for Industrial Purposes. By G. G. ANDRÉ,
F.G.S., Assoc. Inst. C.E. *With* 56 *illustrations and* 12 *plates,* 8vo, cloth,
10s. 6d.

Painting and Painters' Manual: a Book of Facts
for Painters and those who Use or Deal in Paint Materials. By C. L.
CONDIT and J. SCHELLER. *Illustrated,* 8vo, cloth, 10s. 6d.

A Treatise on Ropemaking as practised in public and private Rope-yards, with a Description of the Manufacture, Rules, Tables of Weights, etc., adapted to the Trade, Shipping, Mining, Railways, Builders, etc. By R. CHAPMAN, formerly foreman to Messrs. Huddart and Co.. Limehouse, and late Master Ropemaker to H.M. Dockyard, Deptford. Second edition, 12mo, cloth, 3s.

Laxton's Builders' and Contractors' Tables ; for the use of Engineers, Architects, Surveyors, Builders, Land Agents, and others. Bricklayer, containing 22 tables, with nearly 30,000 calculations. 4to, cloth, 5s.

Laxton's Builders' and Contractors' Tables. Excavator, Earth, Land, Water, and Gas, containing 53 tables, with nearly 24,000 calculations. 4to, cloth, 5s.

Sanitary Engineering: a Guide to the Construction of Works of Sewerage and House Drainage, with Tables for facilitating the calculations of the Engineer. By BALDWIN LATHAM, C.E., M. Inst. C.E., F.G.S., F.M.S., Past-President of the Society of Engineers. Second edition, *with numerous plates and woodcuts*, 8vo, cloth, 1l. 10s.

Screw Cutting Tables for Engineers and Machinists, giving the values of the different trains of Wheels required to produce Screws of any pitch, calculated by Lord Lindsay, M.P., F.R.S., F.R.A.S., etc. Cloth, oblong, 2s.

Screw Cutting Tables, for the use of Mechanical Engineers, showing the proper arrangement of Wheels for cutting the Threads of Screws of any required pitch, with a Table for making the Universal Gas-pipe Threads and Taps. By W. A. MARTIN, Engineer. Second edition, oblong, cloth, 1s., or sewed, 6d.

A Treatise on a Practical Method of Designing Slide-Valve Gears by Simple Geometrical Construction, based upon the principles enunciated in Euclid's Elements, and comprising the various forms of Plain Slide-Valve and Expansion Gearing ; together with Stephenson's, Gooch's, and Allan's Link-Motions, as applied either to reversing or to variable expansion combinations. By EDWARD J. COWLING WELCH, Memb. Inst. Mechanical Engineers. Crown 8vo, cloth, 6s.

Cleaning and Scouring : a Manual for Dyers, Laundresses, and for Domestic Use. By S. CHRISTOPHER. 18mo, sewed, 6d.

A Handbook of House Sanitation ; for the use of all persons seeking a Healthy Home. A reprint of those portions of Mr. Bailey-Denton's Lectures on Sanitary Engineering, given before the School of Military Engineering, which related to the "Dwelling," enlarged and revised by his Son, E. F. BAILEY-DENTON, C.E., B.A. With 140 illustrations, 8vo, cloth, 8s. 6d.

A Glossary of Terms used in Coal Mining. By
WILLIAM STUKELEY GRESLEY, Assoc. Mem. Inst. C.E., F.G.S., Member
of the North of England Institute of Mining Engineers. *Illustrated with
numerous woodcuts and diagrams*, crown 8vo, cloth, 5s.

A Pocket-Book for Boiler Makers and Steam Users,
comprising a variety of useful information for Employer and Workman,
Government Inspectors, Board of Trade Surveyors, Engineers in charge
of Works and Slips, Foremen of Manufactories, and the general Steam-
using Public. By MAURICE JOHN SEXTON. Second edition, royal
32mo, roan, gilt edges, 5s.

The Strains upon Bridge Girders and Roof Trusses,
including the Warren, Lattice, Trellis, Bowstring, and other Forms of
Girders, the Curved Roof, and Simple and Compound Trusses. By
THOS. CARGILL, C.E.B.A.T., C.D., Assoc. Inst. C.E., Member of the
Society of Engineers. *With 64 illustrations, drawn and worked out to scale,*
8vo, cloth, 12s. 6d.

A Practical Treatise on the Steam Engine, containing Plans and Arrangements of Details for Fixed Steam Engines,
with Essays on the Principles involved in Design and Construction. By
ARTHUR RIGG, Engineer, Member of the Society of Engineers and of
the Royal Institution of Great Britain. Demy 4to, *copiously illustrated
with woodcuts and 96 plates*, in one Volume, half-bound morocco, 2l. 2s.;
or cheaper edition, cloth, 25s.

This work is not, in any sense, an elementary treatise, or history of the steam engine, but
is intended to describe examples of Fixed Steam Engines without entering into the wide
domain of locomotive or marine practice. To this end illustrations will be given of the most
recent arrangements of Horizontal, Vertical, Beam, Pumping, Winding, Portable, Semi-
portable, Corliss, Allen, Compound, and other similar Engines, by the most eminent Firms in
Great Britain and America. The laws relating to the action and precautions to be observed
in the construction of the various details, such as Cylinders, Pistons, Piston-rods, Connecting-
rods, Cross-heads, Motion-blocks, Eccentrics, Simple, Expansion, Balanced, and Equilibrium
Slide-valves, and Valve-gearing will be minutely dealt with. In this connection will be found
articles upon the Velocity of Reciprocating Parts and the Mode of Applying the Indicator,
Heat and Expansion of Steam Governors, and the like. It is the writer's desire to draw
illustrations from every possible source, and give only those rules that present practice deems
correct.

Barlow's Tables of Squares, Cubes, Square Roots,
Cube Roots, Reciprocals of all Integer Numbers up to 10,000. Post 8vo,
cloth, 6s.

Camus (M.) Treatise on the Teeth of Wheels, demon-
strating the best forms which can be given to them for the purposes of
Machinery, such as Mill-work and Clock-work, and the art of finding
their numbers. Translated from the French, with details of the present
practice of Millwrights, Engine Makers, and other Machinists, by
ISAAC HAWKINS. Third edition, *with 18 plates*, 8vo, cloth, 5s.

A Practical Treatise on the Science of Land and Engineering Surveying, Levelling, Estimating Quantities, etc., with a general description of the several Instruments required for Surveying, Levelling, Plotting, etc. By H. S. MERRETT. Third edition, 41 *plates with illustrations and tables*, royal 8vo, cloth, 12s. 6d.

PRINCIPAL CONTENTS:

Part 1. Introduction and the Principles of Geometry. Part 2. Land Surveying; comprising General Observations—The Chain—Offsets Surveying by the Chain only—Surveying Hilly Ground—To Survey an Estate or Parish by the Chain only—Surveying with the Theodolite—Mining and Town Surveying—Railroad Surveying—Mapping—Division and Laying out of Land—Observations on Enclosures—Plane Trigonometry. Part 3. Levelling—Simple and Compound Levelling—The Level Book—Parliamentary Plan and Section—Levelling with a Theodolite—Gradients—Wooden Curves—To Lay out a Railway Curve—Setting out Widths. Part 4. Calculating Quantities generally for Estimates—Cuttings and Embankments—Tunnels—Brickwork—Ironwork—Timber Measuring. Part 5. Description and Use of Instruments in Surveying and Plotting—The Improved Dumpy Level—Troughton's Level—The Prismatic Compass—Proportional Compass—Box Sextant—Vernier—Pantagraph—Merrett's Improved Quadrant—Improved Computation Scale—The Diagonal Scale—Straight Edge and Sector. Part 6. Logarithms of Numbers—Logarithmic Sines and Co-Sines, Tangents and Co-Tangents—Natural Sines and Co-Sines—Tables for Earthwork, for Setting out Curves, and for various Calculations, etc., etc., etc.

Saws: the History, Development, Action, Classification, and Comparison of Saws of all kinds. By ROBERT GRIMSHAW. *With 220 illustrations*, 4to, cloth, 12s. 6d.

A Supplement to the above; containing additional practical matter, more especially relating to the forms of Saw Teeth for special material and conditions, and to the behaviour of Saws under particular conditions. *With 120 illustrations*, cloth, 9s.

A Guide for the Electric Testing of Telegraph Cables. By Capt. V. HOSKIŒR, Royal Danish Engineers. *With illustrations*, second edition, crown 8vo, cloth, 4s. 6d.

Laying and Repairing Electric Telegraph Cables. By Capt. V. HOSKIŒR, Royal Danish Engineers. Crown 8vo, cloth, 3s. 6d.

A Pocket-Book of Practical Rules for the Proportions of Modern Engines and Boilers for Land and Marine purposes. By N. P. BURGH. Seventh edition, royal 32mo, roan, 4s. 6d.

The Assayer's Manual: an Abridged Treatise on the Docimastic Examination of Ores and Furnace and other Artificial Products. By BRUNO KERL. Translated by W. T. BRANNT. *With 65 illustrations*, 8vo, cloth, 12s. 6d.

The Steam Engine considered as a Heat Engine: a Treatise on the Theory of the Steam Engine, illustrated by Diagrams, Tables, and Examples from Practice. By JAS. H. COTTERILL, M.A., F.R.S., Professor of Applied Mechanics in the Royal Naval College. 8vo, cloth, 12s. 6d.

Electricity: its Theory, Sources, and Applications.
By J. T. SPRAGUE, M.S.T.E. Second edition, revised and enlarged, *with numerous illustrations*, crown 8vo, cloth, 15s.

The Practice of Hand Turning in Wood, Ivory, Shell,
etc., with Instructions for Turning such Work in Metal as may be required in the Practice of Turning in Wood, Ivory, etc. ; also an Appendix on Ornamental Turning. (A book for beginners.) By FRANCIS CAMPIN. Third edition, *with wood engravings*, crown 8vo, cloth, 6s.
CONTENTS :
On Lathes—Turning Tools—Turning Wood—Drilling—Screw Cutting—Miscellaneous Apparatus and Processes—Turning Particular Forms—Staining—Polishing—Spinning Metals —Materials—Ornamental Turning, etc.

Health and Comfort in House Building, or Ventila-
tion with Warm Air by Self-Acting Suction Power, with Review of the mode of Calculating the Draught in Hot-Air Flues, and with some actual Experiments. By J. DRYSDALE, M.D., and J. W. HAYWARD, M.D. Second edition, with Supplement, *with plates*, demy 8vo, cloth, 7s. 6d.

Treatise on Watchwork, Past and Present. By the
Rev. H. L. NELTHROPP, M.A., F.S.A. *With 32 illustrations*, crown 8vo, cloth, 6s. 6d.
CONTENTS :
Definitions of Words and Terms used in Watchwork—Tools—Time—Historical Summary—On Calculations of the Numbers for Wheels and Pinions ; their Proportional Sizes, Trains, etc.—Of Dial Wheels, or Motion Work—Length of Time of Going without Winding up—The Verge—The Horizontal—The Duplex—The Lever—The Chronometer—Repeating Watches—Keyless Watches—The Pendulum, or Spiral Spring—Compensation—Jewelling of Pivot Holes—Clerkenwell—Fallacies of the Trade—Incapacity of Workmen—How to Choose and Use a Watch, etc.

Notes in Mechanical Engineering. Compiled prin-
cipally for the use of the Students attending the Classes on this subject at the City of London College. By HENRY ADAMS, Mem. Inst. M.E., Mem. Inst. C.E., Mem. Soc. of Engineers. Crown 8vo, cloth, 2s. 6d.

Algebra Self-Taught. By W. P. HIGGS, M.A.,
D.Sc., LL.D., Assoc. Inst. C.E., Author of 'A Handbook of the Differential Calculus,' etc. Second edition, crown 8vo, cloth, 2s. 6d.
CONTENTS :
Symbols and the Signs of Operation—The Equation and the Unknown Quantity— Positive and Negative Quantities—Multiplication—Involution—Exponents—Negative Exponents—Roots, and the Use of Exponents as Logarithms—Logarithms—Tables of Logarithms and Proportionate Parts—Transformation of System of Logarithms—Common Uses of Common Logarithms—Compound Multiplication and the Binomial Theorem—Division, Fractions, and Ratio—Continued Proportion—The Series and the Summation of the Series— Limit of Series—Square and Cube Roots—Equations—List of Formulæ, etc.

Spons' Dictionary of Engineering, Civil, Mechanical,
Military, and Naval; with technical terms in French, German, Italian, and Spanish, 3100 pp., and *nearly 8000 engravings*, in super-royal 8vo, in 8 divisions, 5l. 8s. Complete in 3 vols., cloth, 5l. 5s. Bound in a superior manner, half-morocco, top edge gilt, 3 vols., 6l. 12s.

Crown 8vo, cloth, with illustrations, 5s.

WORKSHOP RECEIPTS,
FIRST SERIES.

By ERNEST SPON.

SYNOPSIS OF CONTENTS.

Besides Receipts relating to the lesser Technological matters and processes, such as the manufacture and use of Stencil Plates, Blacking, Crayons, Paste, Putty, Wax, Size, Alloys, Catgut, Tunbridge Ware, Picture Frame and Architectural Mouldings, Compos, Cameos, and others too numerous to mention.

London: E. & F. N. SPON, 125, Strand.
New York: 35, Murray Street.

WORKSHOP RECEIPTS,

SECOND SERIES.

By ROBERT HALDANE.

SYNOPSIS OF CONTENTS.

Acidimetry and Alkalimetry.
Albumen.
Alcohol.
Alkaloids.
Baking-powders.
Bitters.
Bleaching.
Boiler Incrustations.
Cements and Lutes.
Cleansing.
Confectionery.
Copying.

Disinfectants.
Dyeing, Staining, and Colouring.
Essences.
Extracts.
Fireproofing.
Gelatine, Glue, and Size.
Glycerine.
Gut.
Hydrogen peroxide.
Ink.
Iodine.
Iodoform.

Isinglass.
Ivory substitutes.
Leather.
Luminous bodies.
Magnesia.
Matches.
Paper.
Parchment.
Perchloric acid.
Potassium oxalate.
Preserving.

Pigments, Paint, and Painting: embracing the preparation of *Pigments*, including alumina lakes, blacks (animal, bone, Frankfort, ivory, lamp, sight, soot), blues (antimony, Antwerp, cobalt, cœruleum, Egyptian, manganate, Paris, Péligot, Prussian, smalt, ultramarine), browns (bistre, hinau, sepia, sienna, umber, Vandyke), greens (baryta, Brighton, Brunswick, chrome, cobalt, Douglas, emerald, manganese, mitis, mountain, Prussian, sap, Scheele's, Schweinfurth, titanium, verdigris, zinc), reds (Brazilwood lake, carminated lake, carmine, Cassius purple, cobalt pink, cochineal lake, colcothar, Indian red, madder lake, red chalk, red lead, vermilion), whites (alum, baryta, Chinese, lead sulphate, white lead—by American, Dutch, French, German, Kremnitz, and Pattinson processes, precautions in making, and composition of commercial samples—whiting, Wilkinson's white, zinc white), yellows (chrome, gamboge, Naples, orpiment, realgar, yellow lakes); *Paint* (vehicles, testing oils, driers, grinding, storing, applying, priming, drying, filling, coats, brushes, surface, water-colours, removing smell, discoloration ; miscellaneous paints—cement paint for carton-pierre, copper paint, gold paint, iron paint, lime paints, silicated paints, steatite paint, transparent paints, tungsten paints, window paint, zinc paints); *Painting* (general instructions, proportions of ingredients, measuring paint work ; carriage painting—priming paint, best putty, finishing colour, cause of cracking, mixing the paints, oils, driers, and colours, varnishing, importance of washing vehicles, re-varnishing, how to dry paint ; woodwork painting).

London: E. & F. N. SPON, 125, Strand.
New York: 35, Murray Street.

Crown 8vo, cloth, 480 pages, with 183 illustrations, 5s.

WORKSHOP RECEIPTS,
THIRD SERIES.

By C. G. WARNFORD LOCK.

Uniform with the First and Second Series.

SYNOPSIS OF CONTENTS.

Alloys.	Indium.	Rubidium.
Aluminium.	Iridium.	Ruthenium.
Antimony.	Iron and Steel.	Selenium.
Barium.	Lacquers and Lacquering.	Silver.
Beryllium.	Lanthanum.	Slag.
Bismuth.	Lead.	Sodium.
Cadmium.	Lithium.	Strontium.
Cæsium.	Lubricants.	Tantalum.
Calcium.	Magnesium.	Terbium.
Cerium.	Manganese.	Thallium.
Chromium.	Mercury.	Thorium.
Cobalt.	Mica.	Tin.
Copper.	Molybdenum.	Titanium.
Didymium.	Nickel.	Tungsten.
Electrics.	Niobium.	Uranium.
Enamels and Glazes.	Osmium.	Vanadium.
Erbium.	Palladium.	Yttrium.
Gallium.	Platinum.	Zinc.
Glass.	Potassium.	Zirconium.
Gold.	Rhodium.	Aluminium.

London: E. & F. N. SPON, 125, Strand.
New York: 35, Murray Street.

www.ingramcontent.com/pod-product-compliance
Lightning Source LLC
Chambersburg PA
CBHW021945220326
41599CB00012BA/1189